About The Author

Roy W. Clark was born in Oak Park, Illinois in 1930, raised in middle Tennessee, educated by the United States Air Force, Middle Tennessee State University, Louisiana State University and by life. Shortly after LSU he was a minor coauthor in an "Atoms First" textbook of college chemistry McLellan, Day and Clark, F. A. Davis Company, 1966. After thirty years of teaching chemistry he spent his retirement years reviewing chemistry textbooks and searching for errors in them. He found some and is writing this book about the worst of these errors.

Chemistry: Still Hazy After All These Years

The Disambiguation of Chemistry

Roy W. Clark

Published by New Generation Publishing in 2017

Copyright © Roy W. Clark 2017

First Edition

The author asserts the moral right under the Copyright, Designs and Patents Act 1988 to be identified as the author of this work.

All Rights reserved. No part of this publication may be reproduced, stored in a retrieval system or transmitted, in any form or by any means without the prior consent of the author, nor be otherwise circulated in any form of binding or cover other than that which it is published and without a similar condition being imposed on the subsequent purchaser.

www.newgeneration-publishing.com

Disambiguation refers to the removal of ambiguity by making something clear. **Disambiguation** narrows down the meaning of words and it's a good thing. This word makes sense if you break it down. Dis means "not," ambiguous means "unclear," and the ending -tion makes it a noun.

Dedication

I dedicate this book John Daniel Clark, my brother and my teacher.

I also dedicate this book to William B. Jensen, a **Professor** *at the University of Cincinnati, who has spent his entire career trying to tell his colleagues what is illogical about our teaching efforts, and how we might change them by choosing our words and our definitions more carefully. Thanks Bill.*

Contents

CHAPTER 1 PREFACE 1

CHAPTER 2 THE SMALLEST PIECE 4

CHAPTER 3 CHEMICAL AND PHYSICAL CHANGE 13

CHAPTER 4 THE EMPEROR'S NEW CLOTHES 17

CHAPTER 5 PERIODICALLY CONFUSED 21

CHAPTER 6 DIFFERENT WORLDS NEED DIFFERENT
              WORDS 30

CHAPTER 7 PERIODIC TABLES IN THE TWO WORLDS 37

CHAPTER 8 ELECTROCHEMISTRY 43

CHAPTER 9 RUDY WECKERING 55

CHAPTER 10 CHEMISTRY PHYSICS AND ENERGY 62

CHAPTER 11 FROM ENERGY FIELDS TO ATOMES 66

Chapter 1 PREFACE

"Inevitably there must be a lapse of time between the announcement of scientific advances and their incorporation into elementary courses, but it is unfortunate when the lag is due to the inertia of textbook writers and not because of the inherent difficulties of the concepts."

Laurence S. Foster
Brown University, Sept. 1939

Quoted in *Philosophy of Science* Vol. 37 (1934), p. 157, and in *The Truth of Science: Physical Theories and Reality* (1997) by Roger Gerhard Newton, p. 176.

***"Isolated material particles are abstractions**, their properties being definable and observable only through their interaction with other systems."*
  o "Atomic Physics and the Description of Nature" (1934).
  •Neils Bohr

"What is it that we humans depend on? We depend on our words. Our task is to communicate experience and ideas to others. We must strive continually to extend the scope of our description, but in such a way that our messages do not thereby lose their objective or unambiguous character ... We are suspended in language in such a way that we cannot say what is up and what is down. The word "reality" is also a word, a word which we must learn to use correctly."

Neils Bohr

A preface is supposed to hint to the reader what, in general, this book is about. I taught chemistry for thirty five years. Upon retirement in 1995 I took home a large number of old chemistry textbooks that I and my colleagues had accumulated through the years. Thank you book representatives. As I write this book it is 2016. My hobby in my years of retirement was to look for mistakes in these old textbooks so I could point them out to the new generation of textbook users and writers. The most serious of these mistakes I call blunders. Of course I found many mistakes but I was searching for fundamental mistakes that could not be ignored..... I found one. The blunder I found was this:

The word atom is a polyseme. I know you are shocked at this revelation but it is true.

Polysemy Polysemy is the capacity for a sign (such as a word, phrase, or symbol) to have multiple meanings (that is, multiple semes or sememes and thus multiple senses), usually related by contiguity of meaning within a semantic field. It is thus usually regarded as distinct from homonymy, in which the multiple meanings of a word may be unconnected or unrelated.

*As it is used in chemistry and physics the word **atom** definitely has multiple meanings.* It means both a *lone atom*, not connected to anything else, and it also means an *atom intimately connected* to other atoms by what we call chemical bonds. I repeat:

*As it is used in chemistry and physics the word **atom** definitely has multiple but closely related meanings.* It means both a *lone atom*, not connected to anything else, and it also means an *atom intimately connected* to other atoms by what we call chemical bonds.

The use of the word atom to mean a lone atom and also to mean a bonded atom is a terrible blunder. It tells every student of chemistry that atoms survive bonding with other atoms to form molecules plus or minus energy. The word "molecule" is used to mean the unit formed by the conjoined atoms. This terrible blunder is the polysemy about which I write this book. **I say that an atom bound to other atoms should be continued to be called atom. I say that a lone atom should be renamed to atome to distinguish it from the bound atoms which dominate the study of Earth STP chemistry.**

Fortunately the cure for chemistry's polysemic problem with this atom-of-two-meanings is possible even at this late date. As I see it there are three practical solutions to this problem of the atom's polysemic nature. They are:

1. Rename free atoms to atomes. Leave bonded atoms (conjoined atoms) unchanged.
2. Rename all bonded atom to atomes. Leave free atoms unchanged.
3. Some other creative idea.

I am writing this book advocating solution one. Solution one says the STP world is a world of bonded atoms. The second choice is impractical. The third choice I leave to other chemists who agree that this polysemic blunder is crippling chemistry yet have a better solution than the one I offer. This entire book is a proposal to rename free atoms to atomes, leaving atoms engaged in bonding as they are, atoms. **Atoms involved in bonding to other atoms are still atoms as they have been called since chemistry began. Lone atoms are no longer called atoms but are atomes.**

The following chapters illustrate some of chemistry's history that led up to this polysemic confusion, cures for correction of this atom bound vs.

atom free polyseme, a clarification of what the "periodic table" is a table of. Is it atoms or atomes? I offer a new periodic table (just what we need), and a chapter on electrochemistry (my specialty) in which the term "bonded" means more than "stuck together".

Distinguishing between atoms and atomes will, I hope, clarify many concepts in general chemistry. Chapter five uses the words atomes and atoms to clarify the distinction between types of periodic charts. The chapter on the Clark-Weckering chart of course utilizes the atome vs. atom concept to help authors label their charts as to whether they are macro or atomeic periodic charts. This confusion of the macro and the submicro charts is probably the most confusing of all textbook blunders. Atomes as distinct from atoms should lessen this confusion.

Chapter 2 The Smallest Piece

"But I must confess I am jealous of the term atom; for though it is very easy to talk of atoms, it is very difficult to form a clear idea of their nature, especially when compounded bodies are under consideration".

Michael Faraday 1855
Vol. I Exp. Reserches of the elements.

"Above all, the designation of element should be reserved for the atoms themselves, and for the common forms of the elements the designation of elementary substance should be used."

Rollie J. Myers 2012
Vol. 89 Journal of Chemical Education

"...a man that seeketh precise truth, had need to remember what every name he uses stands for, and to place it accordingly; or else he will find himself entangled in words, as a bird in lime-twigs, the more he struggles, the more belimed."

Thomas Hobbes
Leviathan (p. 56)

"What is it that we humans depend on? We depend on our words. Our task is to communicate experience and ideas to others. We must strive continually to extend the scope of our description, but in such a way that our messages do not thereby lose their objective or unambiguous character ... We are suspended in language in such a way that we cannot say what is up and what is down. The word "reality" is also a word, a word which we must learn to use correctly."

Neils Bohr

When it occurred to people on Earth that they might study things beyond Earth, they utilized what we now call "the laws of physics" to create wonderful instruments called telescopes. These instruments allowed them to dream about what might be going on in these wonderfully unattainable places. Earthlings who did this were called "science fiction writers." More respected Earthlings who indulged in this dreaming exercise were called "astronomers."

These people extrapolated their Earth experiences to far-away places using a great deal of imagination mixed with scientific reasoning to present us with many "pictures" of ugly Martians and of beautiful yet faraway galaxies. The Martians were produced by imagination and not much else. The galaxies were produced by imagination and instruments and sound reasoning processes.

It also occurred to people on Earth to wonder what all the "stuff" surrounding them was made out of. There seemed to be endless kinds of stuff to study so they looked for a few secret ingredients that could be combined to make up anything in our world, but which could never be decomposed into even more fundamental stuff. Earthlings who did this were called "apothecaries." The really smart ones were called "chemists." And the chemists of this nineteenth century were not dumb. They were able to make compounds, analyze them, discover their "formulae" and organize this knowledge into molecular structures to some extent.

They even discovered how to blend "electricity" into some mixtures which occasionally caused elemental substances to be squeezed out one side of their electricity blenders, thus disclosing new and undiscovered elements. Between 1800 and 1900 chemists used these "electrochemical telescopes" to sneak a peek into the world of the very very very small. These instruments allowed them to dream about what might be going on in these unfortunately unattainable places. These people were, to name only a few, Luigi Galvani, Alessandra Volta, Humphry Davy, Michael Faraday and Julius Tafel. These pioneers showed chemists that there was a connection between *electrons* (whatever they were) and chemical reactions (the things the apothecaries were doing). By use of these electrochemical telescopes (we now call them "batteries used for electro deposition"), and even before the electron was understood to be a component of atoms, these pioneers used their electrochemical telescopes to wrench atoms free of their compound prisons, one atom at a time, thus making a *compound substance* into a *simple substance*. I use these italicized words because these words had a meaning to Mendeleev and to his contemporaries that differs from the meaning imparted to twentieth century readers of these words.

I feature Mendeleev and his simple substances in the following with the knowledge that my readers will know the usual and complex history of the search for order in the symbolic chaos with which 19th century chemists struggled. Mendeleev was one of many brilliant chemists, perhaps the best known of this century. Because modern chemists and chemistry teachers rarely use the phrase "simple substance", I pause here to clarify to my readers the difference between a **simple substance**, an **element** and an **atom**. To explain these differences I am including at this point an important quotation from Mendeleev himself on the meaning of chemistry's words. I extend my thanks to William Jensen for making these quotations available.[1]

Begin Mendeleev quotation:

---

[1] Mendeleev on the Periodic Law, William B. Jensen, Dover Publications, 2005.

On the Periodic Regularity of the Chemical Elements

*[Annalen der Chemie und Pharmacie, **1871**, 8 (Supplementband) 133-229]*

Just as the words "molecule," "atom", and "equivalent" were used indiscriminately, one for the other, even as recently as the time of Laurent and Gerhardt, so now the terms "simple substance" and "element" are often confounded with one another. However these terms must be sharply distinguished in order to prevent a confusion of chemical concepts. A simple substance is something material - metal or metalloid - endowed with physical properties and capable of chemical reactions. The idea of a simple substance corresponds to that of a molecule made of one (e.g. Hg, Cd, and possibly other simple substances) or more (e.g., S2, S6, O2, H2, Cl2, P4, etc.) atoms. It is able to display itself in the form of isomeric and polymeric modifications, and is only distinguished from a compound substance by the homogeneity of its material parts. But in opposition to this, the term 'element' designates those material particles of simple and compound substances which determine their behavior from a chemical and a physical point of view. **The word element corresponds to the idea of an atom. Hence carbon is an element, but coal, graphite, and diamond are simple substances.**

End of Mendeleev quotation. I ask you to please reread it until you realize what Mendeleev is saying. Compare your understanding to my understanding. Here is what I read:

A simple substance is something material - metal or metalloid – (a collection of bonded atoms all alike) endowed with physical properties and capable of chemical reactions.

In this sentence he is saying that a simple substance, for example iron, is one giant glob of atoms all of one kind held together by ... he doesn't say. Perhaps chemists of his era of the elements thought gravity held atoms in a pile. If that was what they thought then gases would be single atoms like our inert gases. Liquids and solids would then be condensed atoms and further condense into solids if cold enough. Continuing this speculation on what chemists thought in middle 19th century one can appreciate that Mendeleev' remarks made some sense. If held together (by gravity?) solids were piles of atoms. Liquid were hot piles of atoms, and gases even hotter atom vapour.

This interpretation of the above quotation explains why Mendeleev seemed surprised that in some solid elements there appeared to be allotrope formation (S2 and S6, and also explain his surprise at the diatomic nature of the evaporation products (O2, H2, Cl2, P4). He also tells us the peculiar fact that there are some simple substances which are known to group together in their own private families or" molecules", such as S2, S6, O2, H2, Cl2, P4. These of course were vapors, and his expectations of vaporization would have been the production of monoatomic gases. Then, using carbon as an example he tells us the distinct difference between an element and a simple substance. Mendeleev saw "element" as one unattached atom. He saw bonded atoms, all of one kind, as "simple substances." Simple in that they were groups of atoms all of one kind. Substances in the sense that they were tangible material. He went to great pains to explain that elements (to him "atoms") bonded into a macroscopic bundle might take several forms and his examples were graphite and diamond. Both these simple substances were made of carbon atoms. Since Mendeleev and his contemporaries could not experiment with single atoms they made their periodic charts using the properties of simple substances.

At mid-nineteenth there was no data on single [that is non-bonded] atoms and the periodic charts produced were periodic charts based upon the properties **not of atoms** but of STP elements. This includes Mendeleev's charts, of course. These charts showed some periodicity but the discovery of the noble gases in late nineteenth century showed that what was periodic was not the elements but was the atoms from which they are constructed. Periodicity arose from valence and not from STP properties. Beginning in century 20 chemistry's charts are charts of an atom. They are atomic periodic charts of the atoms.

In chemistry textbooks this is not always made clear. Instead this famous chart is presented in chemistry textbooks as if it were a periodic chart of the STP elements, and later used as what it is, a periodic table of single atoms non-bonded atoms. The reason for emphasis of the STP element table in every textbook is, of course, that this is the level on which the student must work in laboratory. In the classroom the atomic table coexists with the macro table, the table students use in the laboratory. Many books have been written concerning this famous **practical periodic table with little mention that it is fundamentally a table of imaginary elements as they appear on planet Earth.**

The second world of chemistry begins.

Then as the 19th century became the 20th the discovery of single atoms that are not bonded to any other atom at Earth conditions provided us with single atoms with which to work. These were atoms without bonds, and their appearance showed chemists the reason for line breaks in periodic charts. Both chemists and physicists were discovering that the "atom" not 'the smallest piece of the big stuff. Each atom seemed to have an internal construction of their own. Years passed as innovative chemists and physicists poked and punctured these new atoms until they declared that an atom was composed of four things. These "things" were

- Neutrons
- Protons
- Electrons
- Energy

Then they spent several years guessing at what these four things were up to inside the atom. I assume the reader knows the standard story of what they thought they found. Atoms suddenly were no longer the smallest piece of the bigger stuff but instead were tiny little machines constructed of unheard things possessing unheard of properties, and possessing the ability to connect and disconnect from one another. I shall in this book refer to the subsequent imaginings about atoms and their construction as the **second world of chemistry.** The second world of chemistry focused early in the 20th century on explaining

- How protons and neutrons could live together in the nucleus
- What the electrons were doing outside the nucleus? Obviously they were dancing a complex dance in the vicinity of the nucleus.
- Chemists decided they would give the nuclear problems to the physicists, but a chemist called Henry Moseley assigned license plate numbers to nuclei which helped greatly when studying the second word chemical problems of bonding and nonbonding elements and their atoms.

I suppose it is the third world of chemistry. A chemist discovered it, Marie Curie, and the physicists took over from her and created the third world of the vomiting nuclei, popularly known as "radioactivity." Then mid-twentieth century physicists and chemists together created the "great nuclear experiment." They turned nuclei into energy. I conclude that the fourth world consists of nothing but energy. Perhaps before this happens we will figure out what "energy" is. In chapter 10 I review the conundrum "what is energy?"

BUT BACK TO THE FIRST WORLD CHANGING INTO THE SECOND WORLD

As 1890 became 1910 atoms went from the smallest piece to the largest mystery. Atoms were no longer little pieces of the big stuff. They had a delicate architecture of their own which must be discovered. Many designs for the new atoms were proposed from standing still electrons (Lewis theory) to orbiting electrons (Bohr theory). By mid century most chemists had settled on a teachable model of the atom which suggested that the inner electrons in an atom were mysteriously moving while the outer electrons in those same atoms were standing still so as to be easily counted. Chemistry teachers were like the firefly that backed into a fan, delighted no end. They could tell their students about counting the "valence" electrons and ignoring the other electrons in each atom.

Atomic Numbers

In 1916 British physicist by the name of Henry Moseley built a machine to study X-rays. Mosely did not discover X-rays. That had been done in the previous century by Roentgen, who figured out how to generate X-rays. Then Roentgen had used the X-rays to photograph the bones in his wife's hand.

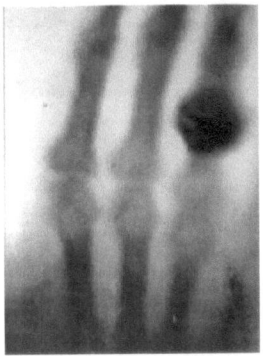

An X-ray was a radiation like light but of very different wavelength, much shorter, and therefore of much higher energy. What Moseley did with Roentgen's waves was to seek a correlation between the wavelengths of the X ray patterns and the metal used as the cathode in his X-ray machine. This enabled him to assign a number to each element, a number which signaled the number of positive charges in the nucleus of the metal used as the cathode. Most textbooks shorten this tale to a simple "Moseley discovered atomic numbers", and he did. I consider this counting of protons as the origin of what I call **the new world of chemistry**. Chemists at last had a technique, to assign a license plate number to each atom, whether alone or bonded. This solved the problem of isotopes and provided numbers for periodic charts to accompany the symbol for an element in the rectangles provided in every periodic table. And so chemists made periodic tables of STP simple substances (which they called elements) and which were bonded atoms and not lone atoms. Since chemical bonding involved the interaction of an atoms with another atom it was reasonable to assume that chemical bonding did not change this identifying number and it was the electrons fooling around with the electrons of nearby atoms that was the essence of "chemical reaction." Clearly in chemical reactions the charge on the nucleus (the atomic number) went unchanged.

Henry Moseley

Here is a photo of Henry Moseley, Did he receive a Nobel Prize for this discovery of the license plate number of every element? Unfortunately he did not, at least not in person, because he enlisted in the British army (WW I) and was killed. He never knew what he had wrought. Nevertheless he lives on in the hearts of all chemists.

Folded Lists: A necessary interlude.

A common misconception of many students is that periodic tables are X Y graphs. They are not. They are **folded lists**. Authors have taken a list of numbers usually horizontally oriented and has folded this list at wisely chosen places so as to convert this list into a "periodic table." All periodic tables in common use are folded lists that differ from each other only in where they are folded and what the symbol arrangement implies.

Mendeleev's 1869 Folded List

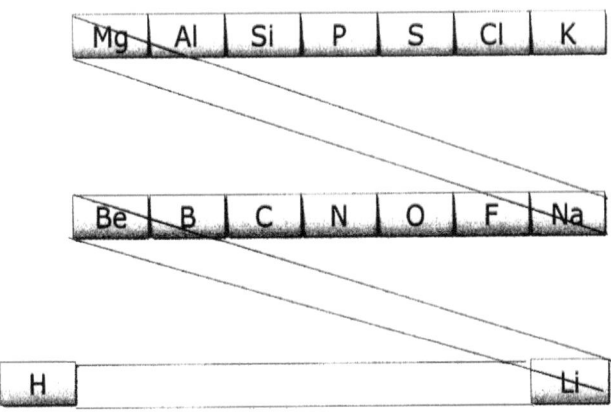

Mendeleev's List Had He Known About Nobility Sooner

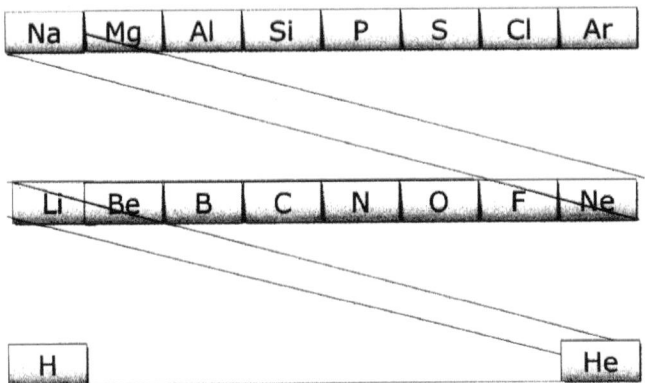

The old examples above include as center data only the symbol of the indicated element. Modern tables are also folded lists, some more folded than others. The drawings above have included as data only the symbol of the element, whereas the current convention is to also include Mosley's identifying atomic number and the atomic weight in amu (atomic mass units). Today's tables other than these primitive examples look vastly different and have center data that tells the symbol chosen for that atom, the atomic number of that atom, the name of the atom (optional), the mass of that atom (in atomic mass units) and **nothing else**. A proper periodic chart is a chart of the known **atomes**. Atomes are unconnected to other atoms at STP. Therefore if the periodic chart is arranged according to the properties of STP elements it comes out wrong.

WHAT'S THAT? Periodic charts are not charts based on elements? They are presented as the symbols of elements but that symbol is ambiguous and can mean two things. It can mean one mole of an element or that same symbol can mean one atom of the chosen element or it can mean can mean one atom of that element. The properties of one mole of the majority of Earth's elements are vastly different from the properties of one atom of that element. It is the *structure of the atomes* that dictates the chemical bonding possibilities built into each atome. The periodic table of elements is very useful to chemists and their students but its arrangement comes from our knowledge of the valences of the atoms symbolized therein.

Therefore the periodic charts we have invented get their shape (where their symbols are folded) from the atom structure. The final structure of periodic charts reveals to us just enough about the element structure for us to be able to assign that a logical place in our periodic table of the elements. I am going to take up this subject in Chapter 5,

Chapter 3 Chemical and Physical Change

"Science has so many dazzling achievements to its credit, we have done so many things which seemed to be impossible, that the popular mind is apt to conclude that, if an explanation is given in the name of science, it must be true whether it be understood or not."

Louis Trenchard More
Author "The Limitations of Science"

"What is it that we humans depend on? We depend on our words. Our task is to communicate experience and ideas to others. We must strive continually to extend the scope of our description, but in such a way that our messages do not thereby lose their objective or unambiguous character ... We are suspended in language in such a way that we cannot say what is up and what is down. The word "reality" is also a word, a word which we must learn to use correctly."

Neils Bohr

The preface introduced the reader to the big blunder, atom bound is atom free. The next most persistent blunder in chemistry textbooks that I have found is the "chemical and physical change" blunder. Like most of the blunders in this book this one originated in the 19th century when chemists knew very little about chemical bonding. They did appreciate that some atoms had the capability of attaching themselves to different atoms to form simple molecules. For the first half of the century they thought metals, and perhaps all simple substances were balls stacked together, no "bonding forces" needed. It followed (they reasoned) that water being composed of these tiny H2O molecules could exist as H2O vapor, H2O liquid, or H2O solid. Just as gold atoms could lie together and make gold so eighteenth century chemists thought that molecules such as H2O could lie together undisturbed and form the liquid and solid phases. The conclusion, because of bonding ignorance, was water forming as H2O was a chemical change because there was a "reaction" between the molecules involved. Textbook writers therefore reasoned that the formation of water molecules by linking H's with an O obviously constituted a chemical change, **and it did**. Water molecules stacked together and becoming unstacked, separated from each other by encountering heat, seemed a different thing entirely and seemed to be simply the physical disengagement of H2Os from other H2Os. This did not deserve the name chemical change, or so they reasoned. As 18tth century authors often said "All of these are H2O so how can this be chemical change?

As the century changed chemists realized that, formula notwithstanding, there were many chemical bonds made and broken in **phase**

changes that had very little to do with the ultimate formula of smallest ratios. Chemical bonds were broken and made in water's phase changes. That makes many of textbooks "physical change" example erroneous, plain wrong.

Since molecule formation was the first obvious necessity for chemical bonding chemists went for many years imagining metal atoms stacked in a pile, all of these "smallest pieces of gold."

Physical as opposed to Chemical Change

When I began teaching general college chemistry in the early 1960s I approved of most of the textbooks of that era with the exception of their opening chapter. This chapter traditionally contained the definitions of science and chemistry-related science words along with the distinctions between the macroscopic view of chemistry and the atomic and molecular view of chemistry. The majority of first chapters devoted several paragraphs to a distinction between those changes involving substances which the student was told to think of as physical change s and other changes in substances which were more properly called chemical changes . My problem as a new teacher was that I never agreed with the logic of this distinction, nor could I see why it needed to be made. Changes were chemical if bonds were broken or formed. Weren't these authors confusing did easy reversibility of some changes with the absence of bond breaking and bond making?

I discussed this with my fellow teachers and found that most of them had no problem with the chemical or physical change distinction, and that only a minority of my fellow teachers wanted to discuss it. I continued to agonize about this for seven years.

Then in 1970 Walter Gensler of Boston University published an excellent critique of this subject in the Journal of Chemical Education[i]. If you wish to fully understand what I am writing in this chapter you must read Gensler's fine paper on the subject of "Terminology Reexamined." Gensler considers the physical change as distinguished from chemical change a fruitless exercise, and sees no reason for its introduction, certainly not at the beginning level.

Since there is a finite chance that the reader, at this juncture, will not stop reading and look up Gensler's paper, I will attempt to relate the core features of his argument here. I wish I could reproduce the entire paper here but copyright restrictions prevent this. Here is my rendition of Gensler's arguments. Remember, this was 1970.

Gensler thinks the distinction between "physical change" and "chemical change" is neither warranted nor wise. Gensler analyses the usual ice-water-water vapor example and concludes it is certainly a chemical change. Gensler suggests "Through first hand experience, everybody knows that, in fact, ice is not water; to maintain otherwise smacks of double talk."

As to the claim that solute into solution followed by recovery is a physical process, Gensler says," So far as recovery from solution is concerned, what is observed is the macroscopic end result of a series of submicroscopic steps, none of which can be defended as "physical."

As to labeling size, shape, state, melting point, temperature, color, density, taste, smell, ductility, viscosity, hardness, thermal and electrical conductivity as physical properties, Gensler has some misgivings about many of these. Especially those properties the detection of which involve breaking chemical bonds, and not only covalent ones.

Gensler concludes with "Since the distinction is not really meaningful to the beginner—in fact, as now presented it is frequently indefensible and confusing—and since the way in which such an empty distinction helps the author in his exposition of chemistry is not obvious, why not forget the traditional distinction and leave it out altogether?"

Then what happened?

After reading Gensler's paper in 1970 I waited patiently for textbooks of chemistry to modify their definition of physical and chemical change to some plausible distinction with better examples, or perhaps skip the whole subject until the student can understand the concept of a variety of chemical bonding concepts involved in seemingly simple changes of state.

I find that forty six years after Gensler very little has changed. New texts still say "Water, steam, and ice are all H_2O. Therefore they are the same substance." Of course I have read only a fraction of the many textbooks created in that forty years, yet in twenty five texts I have reviewed during my retirement I have found only two texts that seemed to take Gensler's advice. The book General Chemistry by Brescia, Arents, Meislich and Turk [ii] ignores the whole idea of physical change as opposed to chemical change, and Chemical Principles by Masterton ,Slowinski and Stanitski[iii] seems to do the same. Nonetheless I think it is fair to say that most texts were not changed by Gensler. Almost all texts still use as an example the water phase changes, claiming the constancy of notation (they are all H_2O) as a demonstration that this was not a chemical change.

In the course of reading chemistry textbooks I discovered that Linus Pauling sidestepped this phase change distinction in the following paragraph:

It is customary to say that under the same external conditions all specimens of a particular substance have the same specific physical properties (density, hardness, color, melting point, crystalline form, etc.}. Sometimes, however, the word substance is used in referring to an element or compound without regard to its state of aggregation; for example ice, liquid water, and water vapour may be referred to as the same substance.

Moreover a specimen containing crystals of rock salt and crystals of table salt may be called a mixture, even though the specimen consists entirely of the one substance sodium chloride. This lack of definiteness in usage seems to cause no confusion in practice.

This avoids "change" and instead blames the confusion on the word "substance." Pauling seems to suggest this is a minor confusion, but it doesn't seem minor to me.

Then in 1997 William B. Jensen in an address to the New England Society, which was fortunately published in the Journal of Chemical Education, addressed this chemical vs. physical change blunder (my opinion) in a gentlemanly and logical fashion by calling it a "logically flawed concept." He concluded that:

"The supposed distinction between chemical and physical changes has formed a part of of the opening chapters of virtually every introductory chemistry text since the late 18th century and you would think that, after 225 years of practice, we would have finally gotten it right."

This blunder lives on in the 21st century.

And now in the days of the internet the physical vs. chemical change definition drags us back into the late nineteenth century. Even should it ever disappear from college texts, I fear it will still be with us until the internet collapses from its own contradictions. Are there not "physical changes?"

Yes there are. In my view physical changes are changes that break no strong chemical bonds yet do involve bending or otherwise stressing these bonds. I offer two examples:

- Vibrating objects such as violin strings are examples of bending bonds without breaking them. If you like you can call this a "physical change."
- Vibrations of atoms within a molecule could, if you like, be called a physical change and would allow physicists to carefully reheat their coffee though they would need a chemist to make it.
- Bending a metal object such as a tuning fork would be a physical change only if it could recover from the stress (no bond breaking) allowed.

Chapter 4 The Emperor's New Clothes

The Emperor of the Elements
A distortion of Hans Christian Andersen's "Keiserens nye Klæder"

Many years ago there was a chemistry teacher so exceedingly fond of studying elements that he spent all his free time arranging element samples in odd patterns on his desk. He cared nothing about reviewing journal articles, going to the theater, or going for a ride in his carriage, except to show off his new elements. He had an element for every hour of the day, and instead of saying, as one might, about any other teacher, "The professor is in a faculty meeting," here they always said "The Professor's in his laboratory."

In the great city where he lived, life was always gay. Every day many strangers came to town, and among them one day came two weavers. They let it be known they were really weavers of theoretical chemistry, and they said they could weave the most magnificent theories imaginable. Not only were their theories uncommonly fine, but lectures delivered using these concepts and their slides had a wonderful way of becoming incomprehensible to anyone who was unfit to be a chemist, or who was mathematically stupid. These arrogant young university students were not actually weavers but instead weavers of magic lantern shows.

Magic lanterns, as you may know, were the slide projectors of the nineteenth century. In the 1820s, the brilliantly focused "limelight," created by igniting oxygen and hydrogen gases on a ball of lime, began to supersede the Argand lantern. Incidentally, this is the origin of "in the limelight," a twentieth century phrase.

These students with their wonderful projector had traveled from a nearby university, and even suggested that they knew more about the new atom than did the esteemed Professor Mendeleev. Although very skeptical, Mendeleev contracted with these weavers to photograph all his element samples, one per slide. Then he paid them an extra sum to finish the show with slides they would create showing pictures of the new theoretical atoms, for he was intensely curious. Remembering their warnings pertaining to invisibility for the unfit he thought to himself, "these slides would be helpful in my classroom for they would show me which of my students were unfit to graduate from my classes."

He paid the slide weavers a large sum of money, whereupon they set up two impressive cameras and proceded to photograph all of his element samples. They then locked themselves into a guest room and pretended to weave, though there was nothing on the looms. Later it was revealed that one of the looms was a notebook computer and the other a Power Point slide projector. All the money which they demanded went into their traveling bags, while they worked the strange instruments far into the night.

"I'd like to know how those atom experts are getting on with the picture show," the Emperor thought, but he felt slightly uncomfortable when he remembered that those who were unfit for their position would not be able to see the show. It couldn't have been that he doubted himself, yet he thought he'd rather send someone else to see how things were going. The whole university knew about the slides' peculiar power, and all were impatient to find out how stupid their fellow teachers were.

"I'll send an honest old colleague to the weavers," the Emperor decided. "He'll be the best one to tell me how the inner atom looks, for he's a sensible man and no one does his duty better." So the honest old colleague went to the room where the two swindlers sat working away at their Power Point presentation. "Heaven help me," he thought as his eyes flew wide open, "I can't see anything at all". But he did not say so. "Don't hesitate to tell us what you think of it," said one of the theoretical experts.

"It is what I always thought atoms looked like. They are beautiful. It is enchanting." The old colleague peered through his spectacles. "Such beautiful atoms." I'll be sure to tell Emperor Mendeleev how delighted I am with it."

"We're pleased to hear that," the swindlers said. They proceeded to name all the orbitals and to explain the M.O.'s, whatever that might be. The old colleague paid the closest attention, so that he could tell it all to the Emperor. And so he did.

The Emperor presently sent another honest old colleague to see how the work progressed and how soon it would be ready. The same thing happened to him. He looked and he looked, but as there was nothing to see in the power point slides they showed him.

"I know I'm not stupid," the man thought, "so it must be that I'm unworthy of my good office. That's strange. I mustn't let anyone find it out, though." So he praised the visions he did not see. He declared he was delighted with the complex orbitals and the antibonding ones too. To the Emperor he said, "It held me spellbound."

All the University was talking of this splendid concept, and the Emperor wanted to see it for himself but he had received such good reports from his colleagues that he decided it to be unnecessary. He set a date for the wonderful slide show.

The classroom was packed. Mendeleev introduced the young weavers and they began the showing of Mendeleev's simple substances. As you will recall Mendeleev called the visible stuff "simple substances" and saved the word "element" for the atom itself. Though most had seen these before the audience applauded warmly after this segment of the show.

Now came slide after slide with nothing on it as one of the weavers named the elements one after the other. Although slides of the lighter elements presented themselves as totally blank finally some of the slides seemed to have a tiny dot centered in the screen. The tiny dots gradually became more visible. The last few slides seemed to have a vague haze surrounding them.

Mendeleev at last stood up and protested that he saw nothing at all! How can the ultimate piece of something be **nothing**? He demanded an explanation.

Then the weavers answered:

Apparently accustomed to this reception of their show, the weavers calmly took turns educating Mendeleev and his students about the new 20th century atom. They told their audience that the old definition of atom was quite unlike the new century atom. The old atoms was a small piece of the old simple substance. Scientists have now discovered that atoms are mostly empty space. Inside the new atom there is almost nothing. Thus the slides show very little. If you have very good vision you saw on the later lantern slides a tiny black dot in the center surrounded by a faint mist. Our teachers tell us that atoms are like a pile of very very very tiny raisins embedded in a cloud of smoke.

RAISINS and a CLOUD of SMOKE

- An atom has a very tiny pile of raisins in its center. Some are red raisins, the others black raisins. Surrounding the raisins (which are very very small) is a cloud of dust. The dust in the dust clouds is supposed to arrange itself in a variety of predetermined patterns but there is some uncertainty that they will obey and actually demonstrate their proper patterns. All atoms are of this nature, the simplest having one red raisin, the next two red raisins and so on. The black raisins were not discovered until this new century and as a result the atomic weights (which included both types of raisin) were often misleading. If Professor Mendeleev had only known this he could have arranged his folded lists differently, ignoring the black raisins.
- The relative size of atoms as calculated by your professor and by Dr. Meyer was very good. However you had no way of knowing how empty these atoms were. Who would have suspected the most of the mass would be in an incredibly dense center?
- In this new century our teachers are working hard to answer two questions. One is "why does the cloud of dust not fall into the raisins?" The other is "Why do some dust clouds bounce off of other

dust clouds, yet others are good friends forever? Perhaps soon we shall discover this.

And with that the weavers took their money and lantern slides and left.

My thanks to:

Jean Hersholt (1886-1956) was a Danish actor who emigrated to the United States, making himself a career in Hollywood as from 1913. He was an avid collector of Andersen editions, and among other things he translated Hans Christian Andersen's fairy tales and stories in the excellent edition The Complete Andersen (six volumes, New York 1949. Further information) - which you may now read on this web site.

By several people, Hersholt's Andersen-translation for the English language world is rated as the standard translation, being one of the best.

http://www.andersen.sdu.dk/vaerk/hersholt/om_e.html
accessed Dec 25, 2012

Chapter 5 Periodically Confused

"The failure of most chemistry teachers to exploit to the fullest the possibilities of modern theoretical principles is illustrated by their resistance on retaining the old Mendeleev form of the periodic chart even though the historical table is obsolete in the light of present day atomic theory....a thorough workable knowledge of the Periodic system and <u>the relationship of this system to electrons of the various atoms</u> is an absolute necessity for any student who aspires to attain any thorough understanding of chemical science."

"What is it that we humans depend on? We depend on our words. Our task is to communicate experience and ideas to others. We must strive continually to extend the scope of our description, but in such a way that our messages do not thereby lose their objective or unambiguous character ... We are suspended in language in such a way that we cannot say what is up and what is down. The word "reality" is also a word, a word which we must learn to use correctly."

Neils Bohr

Harry H. Sislerand Calvin A. VanderwerfJ. Chem. Educ. 1943

From 1890 until 1924 chemistry changed. Before 1890 chemistry was a world of elements at Earth conditions. What were Earth conditions? In truth they were a plethora of different temperatures and also different atmospheric pressures. Properties exhibited by the known elements were temperature and pressure dependent. Chemical reactions of elements with other elements having temperatures and pressures unspecified before and after the experiment were extremely difficult to understand and worst of all were difficult to predict beforehand. Chemistry was a guessing game until practicing chemists (usually on Earth) agreed on a standard atmospheric pressue and a standard temperature (commonly called STP) for reporting chemistry's reaction results. The obvious solution was to choose a "standard temperature and a standard pressure" and to ask the responsible chemist to record the resuts of the experiment accordingly. Had not chemists agreed on this standard for temperature and pressure the teaching and the learning of chemistry would have been impossible. For example when teaching chemistry the H2 plus O2 yields H2O reaction may begin at 25 degrees celsius but it rarely ends at that T and P. It is a rare reaction does not either evolve heat of absorb heat or some form of energy. To complete the reaction the product is brought to STP conditions and the net enegy released or absorbed is recorded.

Behold, the periodic table, or is it tables?

Since the properties of each known element are temperature and pressure dependent any concept of a meaningful "periodic table of the elements" is hopeless without an STP specification. Chemistry on Earth is what chemists do. Chemistry in outer space we leave to the astronomers. The specification of STP by the IUPAC and its use in periodic table creation does not mean that there is no chemistry going on at other temperatures and at other pressures. Of course there is, but this subject will arise in textbooks after the STP periodic table is thoroughly explained.

What is a periodic table? It is a **folded list** of atom symbols accompanied by their atomic number. As I write this that numbered list includes slightly over one hundred different kinds of atom. If they are folded lists there need be some rationalization for the folds. There is. The folding places are chosen by valence considerations. (See Mosley in chapter 2). The method for making this list (of atomic numbers) into a periodic chart of the atoms into a "chart" is simple, or so it seems. Just fold the list of symbols so as to create another portion reside on top of the first list and next fold it again so as to create row three. To illustrate this I reproduce the partial Mendeleev charts illustrated in chapter 2 shortly after atomic numbers were explained.

Mendeleev's 1869 Folded List

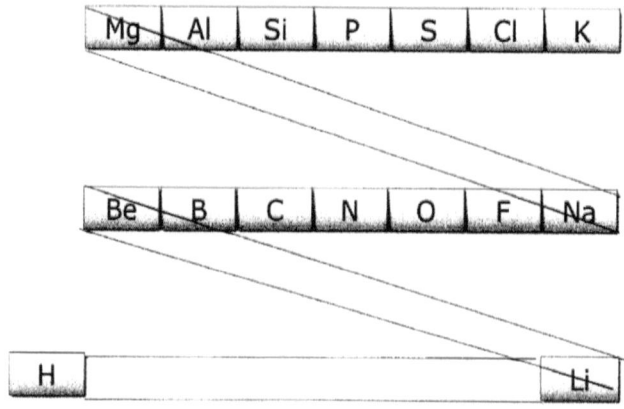

and

Mendeleev's List Had He Known About Nobility Sooner

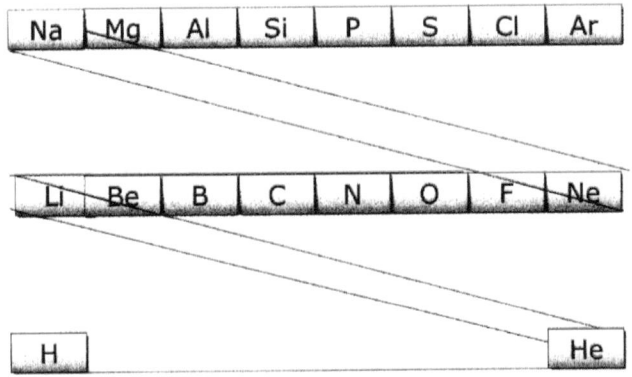

These simplified tables illustrate what I mean when I say that a periodic chart is a folded list. However these folded lists are of little use to the chemist without knowledge of what vertical displacement represents and what horizontal displacement represents. In other words what meaning do we attach to vertical and horizontal displacement and what to the vertical displacements of the "box" and what meaning do we attach to the horizontal displacement of the boxes? What does it mean that some element symbols lie in a vertical column and what does it mean that a new row occurs at each fold? These question must be answered before such periodic tables have any meaning at all.

The answers to these questions about the folds and columns initially is that in folded lists **ordered** by atomic number the list must be **folded** by valence changes. After the noble gases symbols appeared the folding place became obvious. Fold at the right edge of the table leaving yhe family of inert gases as a column along the right edge that were not symbols of atoms. They were and are symbols of the first group of atoms. And of course the reader recalls that atomes are unbonded atoms. So of what are the rest of the symbols in our chart to be called? "atoms?" No. In this chart all the symbols stand for non bonded atoms just waiting for the bliss of bonding.

The unescapable conclusion is that all periodic charts that chemists rely on are charts of atomes at 25 degrees. This explains why there is only one strange column at the right edge of every periodic chart. This column is the never-bonded noble gases which somehow sneaked into our world of ever-bonded atoms. All other columns in the periodic chart are symbols of not-yet-bonded atoms which don't last long in our STP environment.

SUMMARY: Symbols in our periodic charts are symbols of non bonded atoms (with the right-hand exception). If these symbols can somehow escape from the chart and then meet another symbol it is very probable that they will form marriages (even group marriages) and thus form polyatomic molecules. This process is called "chemical reaction."

What are the other symbols in this periodic chart symbols of? this places the zero valent noble gases at the far right where then I repeat: *element symbols were sorted by valence.* 1 for H, zero for He, ,,, **fold**,,,, lithium. Berylium, Boron, Carbon, Nitrogen, Oxygen, Fluorine, Nitrogen, Oxygen, Fluorine, Neon,,,, **fold**, Sodium, Magnesium, Aluminum, Silicon, Phosphorous, Sulfur' Chlorine, argon, ,,, fold, Potassium, Magnesium, and so forth. I think this is far enough into the standard periodic chart to illustrate the general idea of the chart. The x axis of the chart is atomic number (thank you H. G. J. Moseley).The folding sites where valance changes occur and repeats in a cyclic pattern thus causing the next fold are more difficult to comprehend but nevertheless these folding sites occur when a new mode of valance appears. Th folds and boxes of a modern periodic table are to be blamed on valance theory. The modern periodic chart . These symbols are folded by theory. In order to fold a periodic chart one must some theory of the electron structure of each different atom. This need to know the electron structure within each atom in order to design a periodic chart of atoms.

Early chemistry students, and certainly the general public, are not usually aware that the periodic chart is not arranged by the properties of the elements displayed. On the contrary the periodic chart has its strange arrangement as the result of the observed valences of each element that is symbolized. The element valences are displayed by the periodic chart, not their properties. The properties chemists then observe are the result of valence.

In the Mendeleev case the axis to be folded was atomic weight and the purpose was to place vertically above one another elements having similar chemical properties. In the more modern version the noble gases had appeared on the scene and ordered the element symbols more in a more satisfactory manner.

The more atom symbols that were added to such charts the clearer it became that these atom symbols had ordered themselves into columns and rows in patterns that meant similar column members had similar valences. **Valence was the key to the periodic tables.** Atomic weights were not the key to reaction chemistry, combining ability was. Valance and combining ability were the keys to chemistry. Weight just went along for the ride.

In the twentieth century multiple valences were explained by populating each type of atom by electrons (= atomic number) woven into patterns such that similar patterns of this valence type within the atoms placed them in similar vertical columns in the table. This was called matching up the valence types. The result was that the periodic table of atoms became separated into several blocks of symbols (of atomes) in the periodic chart. Because of these complex valence considerations atom symbols (which is what these tables are made of types and it caused chemists with a superior knowledge of valence possibilities to start new columns when valances got difficult to explain. periodic charts into several rows and columns the periodic chart to It was not easy and the results were unsatisfactory using guesses s to where to make the folds.by the 1930s it became quite clear that what these tables were revealing to chemists was that the valence possibilities of each numbered **atom** and their macro grownups called **elements**. The periodic table were created by consideration of the valances observed the **elements.** causing a repeated cycle in this folded list technique. By mid century (20th) the public (and many chemists) were quite content to call this folded list by the name of "periodic table of the elements."

That was fine for writing an illustrated book about simple substances (elements), complete with pictures of STP elements. These books including photos have been very popular with chemistry students and even with the general public. It was not fine for writing a fundamentals of chemistry book. One cannot see electrons, orbitals, or energy. But one can learn the chemists view of these theoretical things useable "things. But there is one aspect all this discussion of atoms that is rather unbelievable to me as a former chemist. It is that so many people believe that the periodic chart is a chart of elements. It is not. It is a chart based on theoretical atome structure.

PERIODIC TABLES OF THE STP ELEMENTS:

Every chemistry textbook is a story of the relationship and connections between periodic tables of visible STP elements and similar appearing periodic tables of the invisible atoms which formed them. Since I am quite old and have read many textbooks of chemistry I can tell you with some confidence that textbooks that display only the single atom type of periodic table are rare. The remainder contain the atomic table **plus** tables of the STP elements. How can one distinguish an atomeic table in a textbook from a macro table of STP elements? Some authors make it easy while others are more subtle about it. Some treat the atomic periodic table of atoms and the periodic table of STP elements as the same thing and illustrate the atomeic table with pictures.

Here is a periodic table. Is it a periodic table of atoms or is it a periodic table of STP elements?

PERIODIC TABLE OF THE ELEMENTS

IA	IIA	IIIB	IVB	VB	VIB	VIIB	VIIIB			IB	IIB	IIIA	IVA	VA	VIA	VIIA	VIIIA
1 H 1.0																	2 He 4.0
3 Li 6.9	4 Be 9.0											5 B 10.8	6 C 12.0	7 N 14.0	8 O 16.0	9 F 19.0	10 Ne 20.2
11 Na 23.0	12 Mg 24.3											13 Al 27.0	14 Si 28.1	15 P 31.0	16 S 32.1	17 Cl 35.5	18 Ar 39.9
19 K 39.1	20 Ca 40.1	21 Sc 45.0	22 Ti 47.9	23 V 50.9	24 Cr 52.0	25 Mn 54.9	26 Fe 55.8	27 Co 58.9	28 Ni 58.7	29 Cu 63.5	30 Zn 65.4	31 Ga 69.7	32 Ge 72.6	33 As 74.9	34 Se 79.0	35 Br 79.9	36 Kr 83.8
37 Rb 85.5	38 Sr 87.6	39 Y 88.9	40 Zr 91.2	41 Nb 92.9	42 Mo 95.9	43 Tc (97)	44 Ru 101.1	45 Rh 102.9	46 Pd 106.7	47 Ag 107.9	48 Cd 112.4	49 In 114.8	50 Sn 118.7	51 Sb 121.8	52 Te 127.6	53 I 126.9	54 Xe 131.3
55 Cs 132.9	56 Ba 137.3	see below 57-71	72 Hf 178.5	73 Ta 180.9	74 W 183.9	75 Re 186.2	76 Os 190.2	77 Ir 192.2	78 Pt 195.1	79 Au 197.0	80 Hg 200.6	81 Tl 204.4	82 Pb 207.2	83 Bi 209.0	84 Po 210	85 At (210)	86 Rn (222)
87 Fr (223)	88 Ra (226)	see below 89-103	104 Rf (267)	105 Db (268)	106 Sg (271)	107 Bh (272)	108 Hs (270)	109 Mt (276)	110 Ds (281)	111 Rg (280)	112 Cn (285)	113 Uut (284)	114 Uuq (289)	115 Uup (288)	116 Uuh (293)	(117) (Uus)	118 Uuo (294)

57 La 138.9	58 Ce 140.1	59 Pr 140.9	60 Nd 144.2	61 Pm (145)	62 Sm 150.4	63 Eu 152.0	64 Gd 157.3	65 Tb 158.9	66 Dy 162.5	67 Ho 164.9	68 Er 167.3	69 Tm 168.9	70 Yb 173.0	71 Lu 175.0
89 Ac (227)	90 Th 232.0	91 Pa 231	92 U 238.0	93 Np (237)	94 Pu (244)	95 Am (243)	96 Cm (247)	97 Bk (247)	98 Cf (251)	99 Es (252)	100 Fm (257)	101 Md (258)	102 No (259)	103 Lw (262)

ANSWER: The answer is that the reader cannot tell. The first two pieces of information are redundant. If one knows the atomic number then they do not need the symbol. The other information is the mass of one atome of this element. Or is it the mass of Avogadro's number of atoms, also known as the molar mass. ? This little piece of information about the table presented table is very important and the student of chemistry must be able to recognize a table of atomes if only in their mind, as distinct from a table of conjoined atoms. Students need to know the atom/molar context of the periodic table just as they need to know the context of what the lecturer is saying. Since this table gives no unit for the mass specifications the viewer cannot answer "Is this an atomice table or an STP table? If this table gave the mass numbers in amu it is an atomeic table. If this same table specified grams per mole for the parenthetic number then the table is a table of STP elements. Once that is cleared up Authors may choose to use the table either way, and often do. In order to make students comfortable with the periodic chart such chartts are complete with pictures of STP elements. Thus the two types are easily distinguishable, atomic or molar.

Clark's suggestion: Give students a helping hand by drawing atomeic table using circles instead of rectangles. Continue to put STP tables in their old familiar rectangles.

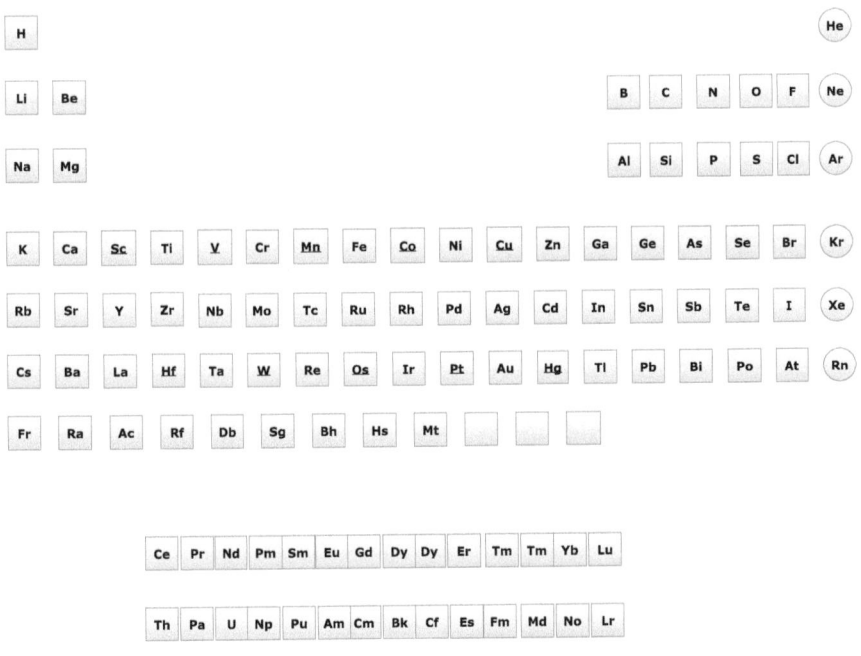

The periodic Table of the STP Elements

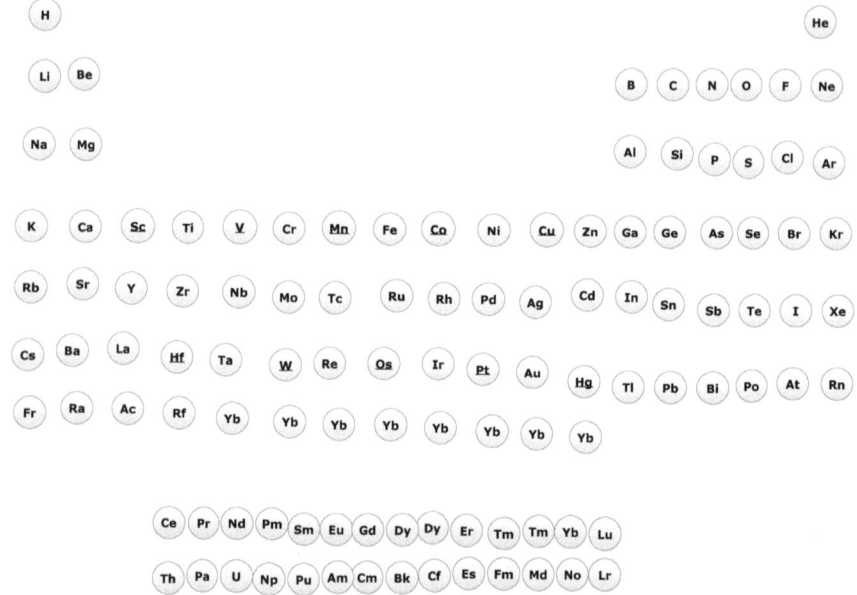

The Periodic Table of Atomes

This convention, if adopted, will make chemistry more easily understood. The Clark suggestion of *atome* equals nonbonded and *atom* equals bonded will fit in with this rectangle vs. circle convention smoothly. Pictures and discussions of STP items will be in a rectangle format. Proposals of periodic tables of the atomes will be in circles. In chapter eight I am proposing a new periodic table I call the Clark-Weckering table. It is an arrangement of symbols in circles. Circles imply the lone atom world while rectangles imply the real world.

PROPERTIES OF THE TABLE OCCUPENTS:

One thing that has always troubled table makers is the question where do I add the list of **properties** to my table? The answer to this question has changed has changed since I began teaching chemistry (1965). Before computers authors had to indicate them somehow into the "box: if there be room. List them outside the table somehow or use color and shading to place property indicators in the tables of elements. After computers there was many methods used to attach properties to each symbol. In today's world computer access is assumed and properties of both atomes and atoms are readily available. I reproduce here a naked table (no indication of lone atomes or STP conjoined atoms), provided with a list of properties via computer.

CHEMIX - PERIODIC TABLE																		

○ Atomic number ○ First ionization potential V ○ Electron configuration
○ Name ○ Specific heat capacity Jg⁻¹K⁻¹ ○ Oxidation states
○ Relative atomic mass u ○ Electrical conductivity ·10⁶ Ohm⁻¹cm⁻¹ ○ Phase 20 °C
○ Melting point °C ○ Thermal conductivity Wcm⁻¹K⁻¹ ○ Crystal structure
○ Boiling point °C ○ Electronegativity Pauling
○ Density g/cm³ ○ Heat of fusion kJ/mol
○ Covalent radius ·10⁻¹⁰m ○ Heat of vaporization kJ/mol
○ Atomic radius ·10⁻¹⁰m ○ Acid-base properties
○ Atomic volume cm³/mol ○ Number of stable isotopes

Group 1/IA	2/IIA											13/IIIA	14/IVA	15/VA	16/VIA	17/VIIA	18/VIIIA
1 H																	2 He
3 Li	4 Be											5 B	6 C	7 N	8 O	9 F	10 Ne
11 Na	12 Mg	3/IIIB	4/IVB	5/VB	6/VIB	7/VIIB	8/VIII	9/VIII	10/VIII	11/IB	12/IIB	13 Al	14 Si	15 P	16 S	17 Cl	18 Ar
19 K	20 Ca	21 Sc	22 Ti	23 V	24 Cr	25 Mn	26 Fe	27 Co	28 Ni	29 Cu	30 Zn	31 Ga	32 Ge	33 As	34 Se	35 Br	36 Kr
37 Rb	38 Sr	39 Y	40 Zr	41 Nb	42 Mo	43 Tc	44 Ru	45 Rh	46 Pd	47 Ag	48 Cd	49 In	50 Sn	51 Sb	52 Te	53 I	54 Xe
55 Cs	56 Ba	57 La	72 Hf	73 Ta	74 W	75 Re	76 Os	77 Ir	78 Pt	79 Au	80 Hg	81 Tl	82 Pb	83 Bi	84 Po	85 At	86 Rn
87 Fr	88 Ra	89 Ac															

Lanthanides →	58 Ce	59 Pr	60 Nd	61 Pm	62 Sm	63 Eu	64 Gd	65 Tb	66 Dy	67 Ho	68 Er	69 Tm	70 Yb	71 Lu
Actinides →	90 Th	91 Pa	92 U	93 Np	94 Pu	95 Am	96 Cm	97 Bk	98 Cf	99 Es	100 Fm	101 Md	102 No	103 Lr

This periodic table tries to be two periodic tables, both the elemental table and the atomeic table in one rectangular setting. Properties, however, are a very different matter for elements than for atoms. This and many other ambiguities are the reason for this book. In this book I shall lessen this confusion by the use of charts within rectangles in the charts of macro elements, and the use of circles in charts of single atom visions and behavior. Notice that matter here is used in one of its senses for matter also is a polyseme.

Box or circle? It matters.

Chapter 6 Different Worlds Need Different Words

Polysemy   Polysemy is the capacity for a sign (such as a word, phrase, or symbol) to have multiple meanings (that is, multiple semes or sememes and thus multiple senses), usually related by contiguity of meaning within a semantic field. It is thus usually regarded as distinct from homonymy, in which the multiple meanings of a word may be unconnected or unrelated.

Disambiguation refers to the removal of ambiguity by making something clear. Disambiguation narrows down the meaning of words and it's a good thing. This word makes sense if you break it down. Dis means "not," ambiguous means "unclear," and the ending -tion makes it a noun.

My parody on "Emperor" was intended to remind readers that chemistry is now conceptually two different worlds. The 19th century words of chemistry which sufficed in the old world are not always appropriate when concepts fall into the new and very tiny world. The resulting struggle to define chemistry's second world, the world of lone atoms, was very encouraging to chemists because they could at last explain why some atoms collided and joined while others collided and did not join. In studying this largely imaginary world of theoretical atoms chemist apparently words so that they are useful in both worlds has not been entirely successful. In particular the use of the word atom to mean both a lone atom and a bound atom has, in my opinion, been chemistry's worst blunder. It is a blunder because the new lone atoms (that appeared at the end of the 19th century called noble gases) were taken as verification that elements were "made out of atoms." Therefore it followed that elements consisted of conjoined atoms, and what conjoined them was anybody's guess.

Of course it was not a blunder to discover lone atoms, but it was a horrible blunder to imagine a molecule as simply a string of atoms. What would cause a string of atoms to stick together and not fall apart? Well chemists had already answered some of these questions with theories of "chemical bond formation." What are bonds? They are thought to be an explanation of why colliding atoms might stick together and thus form atom groups which are then called molecules. Bonds then are the reason some atoms decide to get married while others refuse. Atoms formed molecules because they became bonded. If they did not bond they rebounded and went looking for a different atom to which they might bond. What chemists wanted to know about this bonding or not bonding was which atoms would bond and the circumstances that might promote desired bonding between atoms.

But what caused bonding between two atoms? According to Dalton's atomic theory two circles (the atoms) sat together under the right conditions thus forming molecules. Dalton, of course, had no concept of why atoms might stick together and so he drew then as adjacent circles.

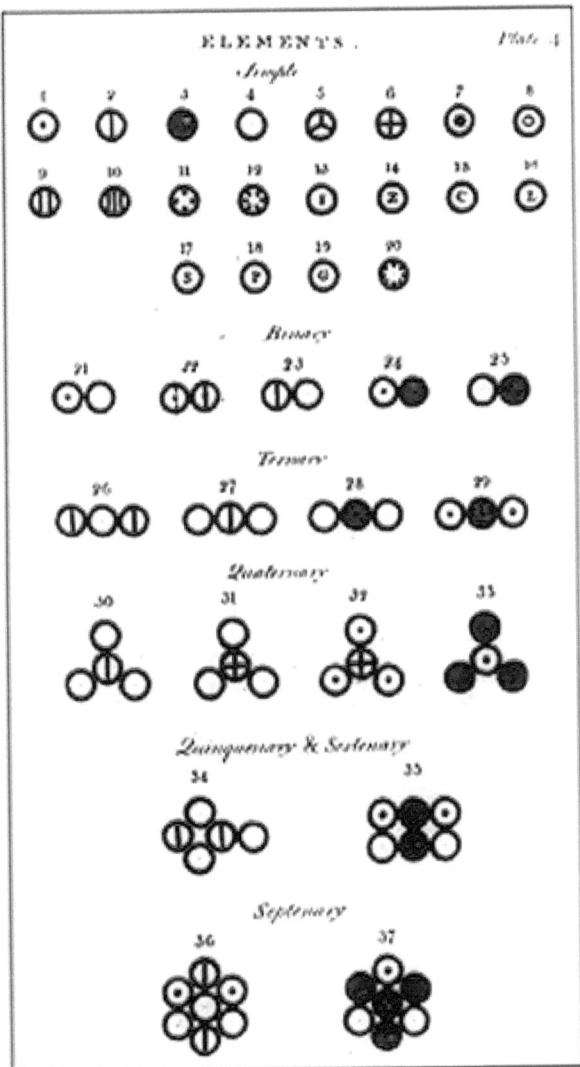

John Dalton's *A New System of Chemical Philosophy*

These drawings were important. They suggested that atoms might link together forming molecules, but why would some atoms link and others ignore each other? And for the one's that might link **how** would atoms link. Surely not just by touching each other.

Well a century passed and chemists studied what atoms might link with other atoms to form the multitude of compounds that were observed. The problem of what might link the atoms to each other and hold them in a molecule was a difficult one. Chemists finally resorted to a dash to represent the chemical bond and let other chemists worry about the details of bonding. The dash or hyphen would do the job of indicating the information that the two atom binding had been accomplished and that the prior lone atoms were now two atoms firmly stuck together. Eventually chemists figured out what was going on when two atoms bonded with each other thus becoming, well "bonded atoms." Bonding, it turned out was an essential part of molecule formation. It was not sufficient for atoms to simply collide. With the help of energy the two atoms involved will be distorted so as to form a bonded pair. I say again that the two atoms with the help of energy will be distorted so as to form a bonded pair.

What is so often overlooked is what that energy gain or loss does to the atoms involved in the bonding. I can tell you what it does to the atoms involved. This energy distorts the atoms involved. The atom before bonding is not the same as the atom after bonding. I hope the reader recognizes that this statement is the raison d'etra for this book. The word atom is a polyseme. **Atom bonded is not atom free.** I contend that this polyseme be resolved by renaming free atom to atome. Bonded atoms shall remain atoms.

Is this an unrealistic idea? No it is not. I have written this book showing how this new spelling of one word could make a large diference to the teacher and student in chemical education. It will also unmuddy all fields of science to their advantage.

THE TEACHING OF THE ATOM OR ATOME CONCEPT

As a former teacher I feel confident that the transition from "atoms" for everything to "atoms" for bonded atoms while using atomes for or no will be easy to teach. I offer two examples of teaching aids for this concept. One is a picture intended to illustrate the distinction between atoms, atomes , and atome fragments.

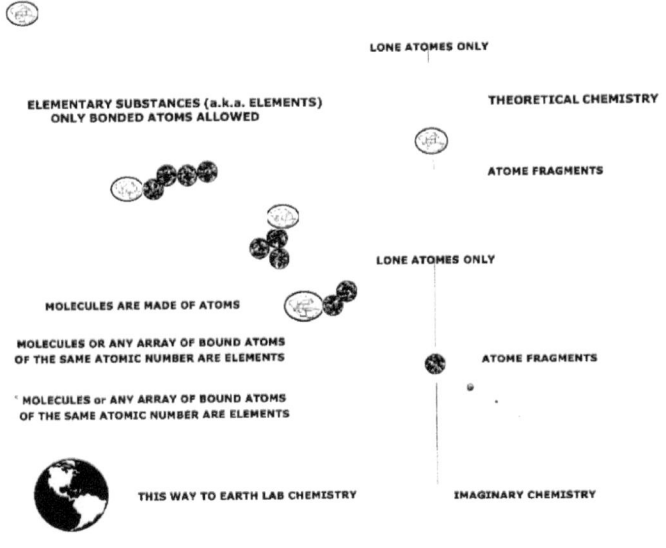

This diagram says in one picture what has taken me many pages to explain. I hope it is useful.

My second teaching aid is a diagram that appeared in a journal article by William Jensen (cite reference here.)
William B. JensenEd.chem.wisc.edu. Vol 75 No. 7 July 1998
J. Chem

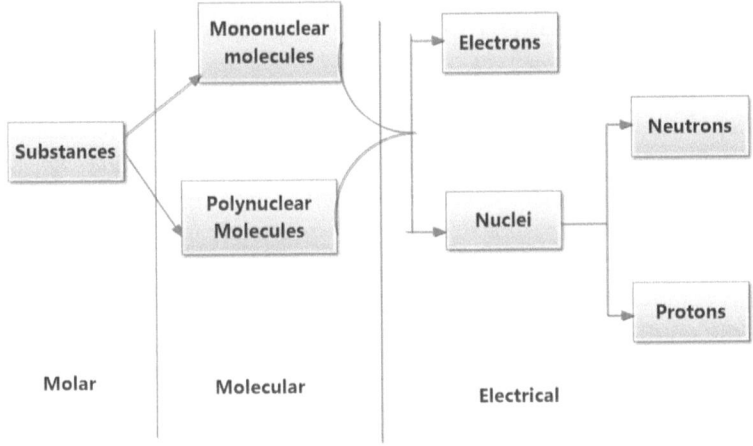

Jensen calls this a corrected composition/structure hierarchy. If read in the direction of the arrows this is an "are made out of" diagram. If read in the opposite direction it is a "what they are made out of "diagram. Applying this diagram to Earth chemistry the student of chemistry discovers a shocking thing. Mononuclear molecules are rare! This is why they are referred to as "the rare gases." The conclusion of most chemists is that these rare gases will not bond to any other elements. On the diagram these gases are labeled "mononuclear molecules." The alternate is called "polynuclear molecules" which means that atoms have bonded together to form multi-atom units. This is the usual case at STP so that "polynuclear molecules" means bonded species composed of two or more atoms that are bonded.

Currently chemists use the name atom for bonded atom and for bonded atom. Thus the word atom has two meanings. As you now know, I propose that atom be changed to atome for any nonbonded atom. This will make noble gases atomes and leave bonded atoms to be atoms as they have been for centuries, atoms. I choose to leave bonded atoms as atoms and to make free nonbonded atoms atome(s). Mononuclear molecules are atomes. Polynuclear molecules contain atoms as they have for centuries.

My third teaching aid is a quotation from an organic synthesis chemist.

MY CONVENTION AND SYNTHETIC CHEMISTRY:

This quotation is from a chemist whose specialty was synthetic chemistry. Here is what he told his students.

Old Atoms, New Atoms, Single Atoms and Married Atoms

"Why can the chemist not take the requisite numbers of atoms and simply put them together? The answer is that the chemist never has atoms at his disposal, and if he had, the direct combination of the appropriate numbers of atoms would lead only to a Brobdingnagian potpourri of different kinds of molecules, having a vast array of different structures. What the chemist has at hand always consists of substances, themselves made up of molecules, containing defined numbers of atoms in ordered arrangements. Consequently, in order to synthesize any one substance, his task is that of combining, modifying, transforming, and tailoring known substances, until the total effect of his manipulations is the conversion of one or more forms of matter into another."

Robert Burns Woodward
USA chemist

Now I shall restate Woodward's words using atomes for naked atoms.
"Why can the chemist not take the requisite numbers of atomes and simply

put them together? The answer is that the chemist never has atomes at his disposal, and if he had, the direct combination of the appropriate numbers of atomes would lead only to a Brobdingnagian potpourri of different kinds of molecules, having a vast array of different structures. What the chemist has at hand always consists of substances, themselves made up of molecules, containing defined numbers of atoms in ordered arrangements. Consequently, in order to synthesize any one substance, his task is that of combining, modifying, transforming, and tailoring known substances, until the total effect of his manipulations is the conversion of one or more forms of matter into another."

Robert Burns Woodward
USA chemist

MY ATOME AND ATOM IDEA AND THE IUPAC:

What will IUPAC think about all this atome idea? IUPAC, the International Union of Pure and Applied Chemistry has long been in charge of defining chemistry's words. If chemistry should accept my concept of atome should be the name for free atoms and that atom should be the name for bound atoms then "atom" will be redefined accordingly. That will not be difficult for an atome will be changed considerably by the contortions of the bonding process. Atoms are much changed by bonding and the changes they undergo are already described by theories involving waves and or theories involving particles (electrons). Also IUPAC will have to redefine chemical element to accommodate the concept that nonbonded atomes are quite changed by bonding to form atoms. Here is the present current definition of element.

IUPAC
chemical element

1. A species of atoms; all atoms with the same number of protons in the atomic nucleus.
2. A pure chemical substance composed of atoms with the same number of protons in the atomic nucleus. Sometimes this concept is called the elementary substance as distinct from the chemical element as defined under 1, but mostly the term chemical element is used for both concepts.

END OF DEFINITION.

Clark, comment on the IUPAC definition:
Part one of the IUPAC definition of "element" is based upon the atomic number with absolutely no regard for the atoms themselves. Elements are

made from atoms whether lone atoms or bound atoms. Whatever state the atom's structure may be in, bonded or not, if the pieces all have the same atomic number it is declared and element. There is no regard of the electrons at all in this definition at all. This definition is a statement that atoms are atoms regardless of bonding or no.

The second definition is a hint that elements when bonded are composed of atoms also, but these atoms are distorted by the bonding such that both should not be called atoms. This is of course my concept that there are two species of atom, one free and lonesome, the other bonded to one or more atoms. Furthermore atoms free are not atoms bonded. Bonding changes atoms.

This polysemous use of atom is, of course, the reason for this book. I advocate atome for lone atom and atom for bonded atoms.

Chapter 7 Periodic Tables in the Two Worlds

'What is the use of a book', thought Alice, 'without pictures or conversations?'

Lewis Carroll
Alice's Adventures (1865).in Wonderland

Thank you Alice. I shall begin with a picture and follow it with conversation. Here is the first picture.

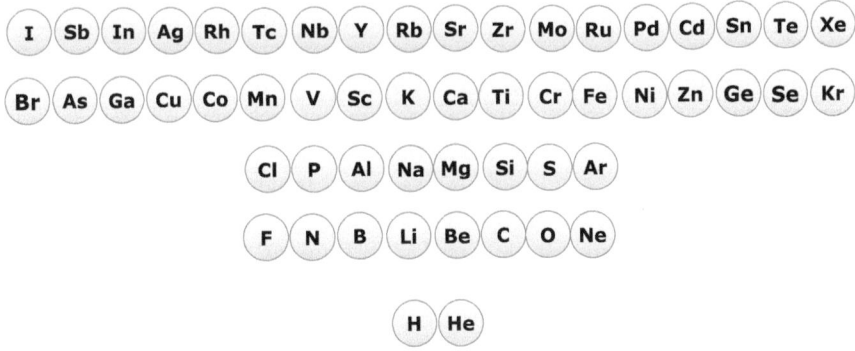

This is the first five rows of the Clark-Weckering Periodic Table. If you are Alice in general chemistry instead of Alice in Wonderland you will readily see that it is not a folded list but instead is two juxtaposed folded lists set side by side.

The Clark-Weckering Periodic Table

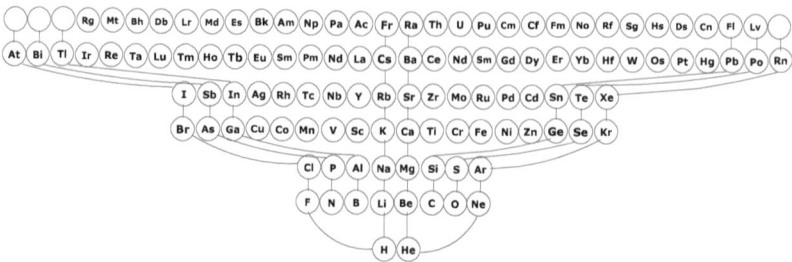

This is the complete Clark-Weckering periodic table. It is called the Clark-Weckering table because I (I'm Clark) was inspired by an unknow Belgian Engineer named Rudy Weckering. In my career as a teacher I happened across a book by Rudolph Weckering. In this 1935 book offered an extension of G. N. Lewis' ideas of a static electron structure for electrons. Weckering called his new pattern of nodes on which electrons might reside the nodic field. Weckering constructed excellent seemingly 3D pictures of histhat "field" and dozens of three dimensional molecule drawings which he claimed would solve all chemical bonding problems. I was impressed by Weckering's ideas but his contemporaries paid him little heed. Rather than tell you further of Rudy I devote a separate chapter to Weckering and his ideas. I attach his name to mine on the table I have drawn so that chemists will remenber this ingenious Belgian engineer, as he called himself.

Back to the C-W table: The table is not a folded list as are other tables. It instead is two folded lists, one of the odd-numbered elements and the second of even numbered elements. Combining this with the "long form periodic tables of the past results in a built-in symmetry which I believe more clearly presents the vertical relationships and emphasies the diagonal relationships that are disguised by the usual tables. Two examples will suffice. The alkali metals show a linear relationship whereas the halogens show a diagonal relationship. Another feature of this table is that the mysterious suscripted elements align smoothly with the normal elements.

The bald truth is that the IUPAC design owes its popularity not to its chemical logic so much as it ability to fit in one page of a textbook.

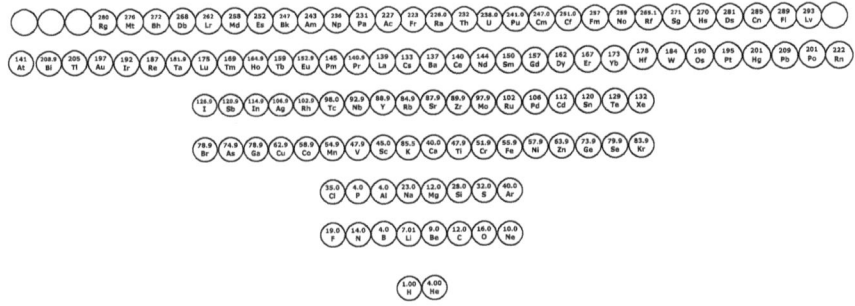

The Clark- Weckering Periodic Table
with some connecting lines

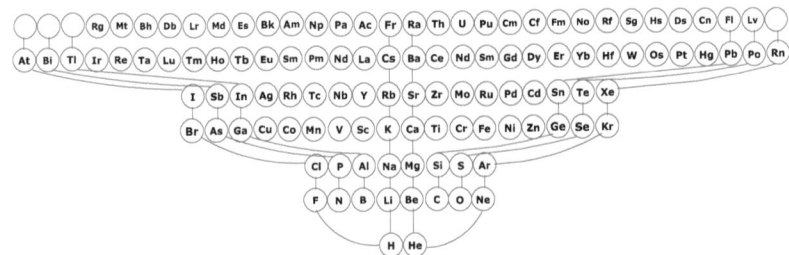

Why did I show the shortened version first? Because if one is looking for relationships in the C-W table they are there but perhaps more difficult to see. So here is a version with the reationships connectrd by lines to make these attributes of the table more obvious.

If you are not too stunned by the beauty of this table you are now asking where are the vertical, horizontal and diagonal relationships found on this table of elementes. They are there but more of them are diagonal relationships and fewer are vertical. For example the alkali metals are along a diagonal as are the noble gases. The alkali metals and the alkaline earths are standing up straight as is the custom. This extensive diagonalism may bother some introductory chemistry students and so I have prepared a beginners Clark-Weckering version to aleviate this problem. Here is the alternate version which emphacises the vertical associations. It is the same table but easier for beginners.

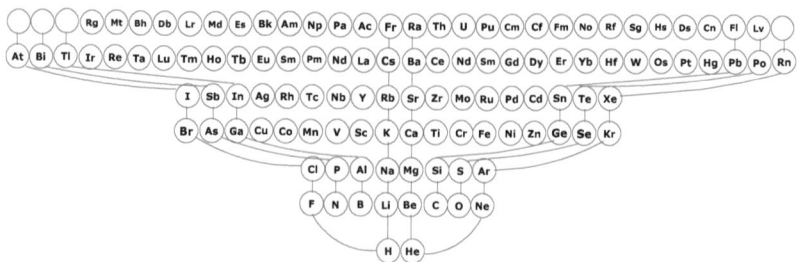

And here is another version for beginners which I call "The table which looks like a table.

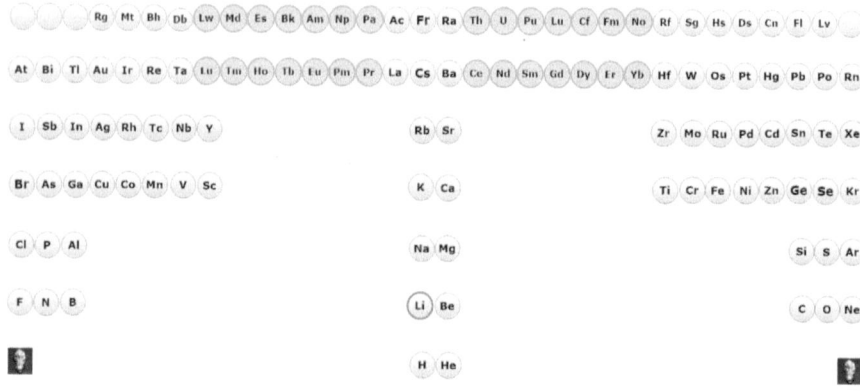

This table can be thought of an unfolded C-W table with two fake feet added to make it look like a table. Students will like the vertical associations. Mature chemists will prefer the beautiful triangular version.

Ah but there is even one more version of the C-W table this one for advanced students and instructors. In mid century it became obvious that thinking of these horizontal placements as descriptions of the "energy levels" of the seven rows in the table. By mid twentieth century in the atome, calculations from theory showed that as electrons were added to atoms they did not simply occupy one of seven levels. Instead, as more electrons joined the resident electrons to make a new elemente, the energy levels shifted according to sublevel with the result that the main energy levels were fragmented into these sublevels. So it came to pass that in the last quarter of the 20th century tables were drawn up to visualize the splitting of d-levels from p levels and so forth. Longuet Higgins published this advanced sublevel energy shift table in the 1960s and I have simply changed it to the odd-left even-right C-E form. It is the best form for chemists but I do not expect it to become popular with the general public.

Why?

Because all C-W tables are tables of atoms, not tables of elements.

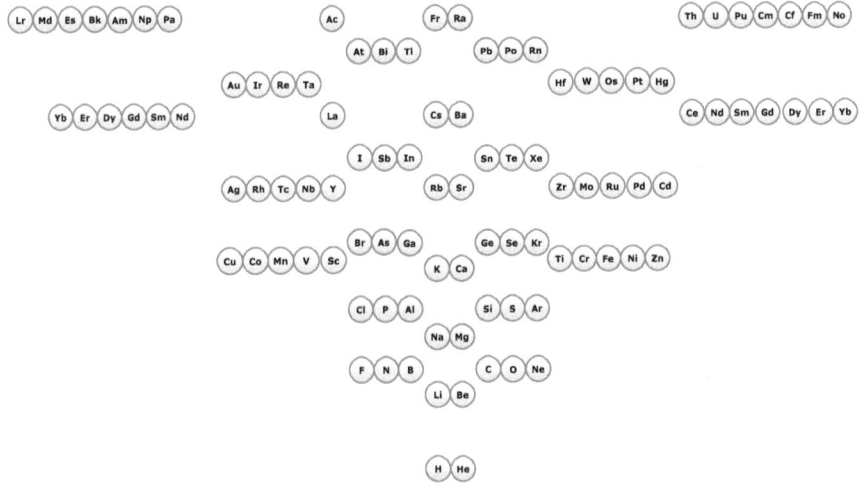

This version of Longuet's-Higgins' table I have split into odd-even portions consistent with the Clark Weckering Odd even separation.

"Periodic table" is a Polyseme

" . . . a man that seeketh precise truth, had need to remember what every name he uses stands   for; and to place it accordingly; or else he will find himself entangled in words, as a bird in lime- twigs; the more he struggles, the more be belimed."

Thomas Hobbes (1588-1679)
English philosopher

Here is a multiple choice question for the chemistry teachers who may read this book. does the text from which you currently teach have a flyleaf periodic table? If not does it have prominent page-consuming table for student use?

1. Is this table a periodic table of atoms only?
2. Is this table a periodic table of elements at STP only?
3. Is this table a combination table which is intended to be used for both elements and atoms, expecting the student to choose the correct context?
4. Is the table colored or shaded in any way so as to mark off metals from nonmetals?
5. Are your students confused as to which table is which?

I would guess that it is up to you to distinguish clearly that the actual

(IUPC approved) periodic table is a table of lone atoms, atoms not connected to any other atom. This is what should be made clear at the start of all chemistry textbook. When you go to your laboratory this atomic table (at first) will do you little good because everything in your laboratory is polynuclear. All the substances you will work with will be polynuclear with one possible exception. You might be assigned to work with helium, a noble gas.

So here is the all-purpose table in everybody's textbook. I have modified it only slightly to show that, STP or not, the last column represents monoatomic gases, that is atoms.

1 H 1.0																	He
3 Li 6.9	4 Be 9.0										5 B 10.8	6 C 12.0	7 N 14.0	8 O 16.0	9 F 19.0		Ne
11 Na 23.0	12 Mg 24.3										13 Al 27.0	14 Si 28.1	15 P 9	16 S 32.1	17 Cl 35.5		Ar
19 K 39.1	20 Ca 40.1	21 Sc 45.0	22 Ti 47.9	23 V 50.9	24 Cr 52.0	25 Mn 54.9	26 Fe 55.9	27 Co 58.9	28 Ni 58.7	29 Cu 63.5	30 Zn 65.4	31 Ga 69.7	32 Ge 72.6	33 As 74.9	34 Se 79.0	35 Br 79.9	Kr
37 Rb 85.5	38 Sr 87.6	39 Y 88.9	40 Zr 91.2	41 Nb 92.9	42 Mo 95.9	43 Tc 99	44 Ru 101.0	45 Rh 102.9	46 Pd 106.4	47 Ag 107.9	48 Cd 112.4	49 In 114.8	50 Sn 118.7	51 Sb 121.8	52 Te 127.6	53 I 126.9	Xe
55 Cs 132.9	56 Ba 137.4	57-71	72 Hf 178.5	73 Ta 181.0	74 W 183.9	75 Re 186.2	76 Os 190.2	77 Ir 192.2	78 Pt 195.1	79 Au 197.0	80 Hg 200.6	81 Tl 204.4	82 Pb 207.2	83 Bi 209.0	84 Po 210.0	85 At 210.0	Rn
87 Fr 223.0	88 Ra 226.0	89-92															

57 La 138.9	58 Ce 140.1	59 Pr 140.9	60 Nd 144.2	61 Pm 147.0	62 Sm 150.4	63 Eu 152.0	64 Gd 157.3	65 Tb 158.9	66 Dy 162.5	67 Ho 164.9	68 Er 167.3	69 Tm 168.9	70 Yb 173.0	71 Lu 175.0
89 Ac 132.9	90 Th 232.0	91 Pa 231.0	92 U 238.0											

This is **not** the new Clark-Weckering table I designed to sort out the chaos of information contained in today's "periodic tables." No. This is the "standard" table everyone uses because their teacher told them to. It is in all textbooks Students need this periodic table for the information that is often piled into it, everything from average molar mass to STP properties. It is left up to the student to guess which of the listed properties are "avogadros number of atoms" and which pertain to a lone atom (which I call an atome). I propose that STP element tables should consistently be displayed in rectangles ever be displayed inrectangles like this table is.Thenperiodic tabes of lone atoms shoud utilize circles not rectangles. The design of both tables is based on atomes and not on atoms.

Moer tables:

Chapter 8 Electrochemistry

"Knowledge has to be sucked into the brain, not pushed into it."

Victor Weisskopf
American physicist

I am about to explain to my readers a bit of electrochemistry. I know what you are thinking. All of my readers already understand electrochemistry because they teach it yearly to students, most of whom are **barely** familiar with some of electrochemistry's key words and thus grasp at vague ideas like "lots of volts are bad but little volts are okay." And "volts can be stuffed into batteries and come out when you call them." I decided to offer here the Roy Clark explanation of electrochemical cells which I hope sucks (see quote above). The very word "electrochemistry" implies the movement of charged particles through chemicals, but what are these charged particles and under what circumstances do they move? Why don't they just stay home?

The two types of charged particle which can be made to move and with which electrochemistry is concerned are **electrons** and **ions**. Electrochemistry is (in part) the science of using our knowledge of ion motion in solution (slow) with our knowledge of the electron's chemical potential energy difference between unlike metals in order to create an "electrochemical cell." This electrochemical cell after its completion

Chemists have found that electron conduction with charged ion conduction. These are the two major mechanisms by which "charges" are compelled to move not randomly but in an ordered fashion. Thus I begin with two definitions, of electrical conduction. These are "conductors of the first kind" and "conductors of the second kind."

"**Conductors of the first kind**" are solids or liquids with a special characteristic. The atoms of which they are composed have formed chemical bonds with their neighbours holding them together. This of course is what makes them solids or liquids. But these solids which we call **metals** all possess what I call "friendly valence electrons." These are outer electrons are in an electron environment such that that, if the proper electric field comes along, they love to travel and thereby visit their neighbouring atoms. Physicists call them "conduction electrons" and blame them for metals and alloys' property of easily moving those tiny little critters happily through wires. Metals and alloys conduct electricity in this way. Electrons move. Everybody else stays home. The result is that conduction of the first kind is much faster than conduction of the second kind.

What are **conductors of the second kind**? Well they are things that conduct electricity by **ion movements** as opposed to **electron movements**. Salt water is an example of an electrolyte solution. Solid salt is not an electrical conductor at all, Pure water is a very poor conductor of electricity for reasons you probably have learned. A cube of salt may be composed of ions but the ions cannot move and cubic salts are nonconductors. However chemists know that "There is no salt in the sea." Only ions survive that plunge at the beach. Therefore seawater is a "good conductor of the second kind as is any aqueous solution of electrolytes. A second example of a conductor of the second kind is a molten salt such as molten NaCl without any solvent. Ionic solutions and molten salts conduct electricity because the **ions** can move and in fact are always in motion due to thermal energy. Since ions in solution are bound up with water molecules they do **not** move very rapidly. As we shall see shortly this slow motion is the very reason why "batteries" work. If ions in a solution moved as rapidly as electrons do in a metal your cell phone would last about five minutes before it needed recharging. Ions in circuits are traffic controllers.

Electrochemical cells (often called "batteries" by the unwashed) are simply an ingeneous arrange ment of conductors of the first kind with conductors of the second kind This ingeneous arrangement most often involves the use of two different metals which are seperated from each other by ionic solutions. There are electrochemical cells which utilize only one metal for both electrodes although this is rare. The case of identical electrodes will be explained later. For now we assume that the right electrode is a **different metal** than is the left electrode,certainly the most usual case.

REACTING ZINC with COPPER Solution.:

Imagine you have made a solution of some soluble copper salt in water, say the sulfate of copper.What does it contain? Hydrated (it is in water) copper ions and hydrated sulfate ions. Now you go to the stockroom and talk them out of a strip of zinc. You go back to your lab and immerse the zinc into the copper solution. What happens? Copper metal begins to plate out on the strip of zinc Shocking as this may be to you, chemists have known since the 18th century that some metals will go into solution and kick out of solution some inferior metals in the process. Zinc will replace copper from its solutions.

Similar experiments with other solutions and other metals permits chemists to make a list called the electromotive series (It is in your textbook). Further in this lecture you will see that zinc will replace silver with even more shocking results. Before continuiing we must ask and answer a crucial question: "What does all this dipping of one metal into a solution of another have to do with the previous talk about conductors of the first kind and conductors of the second kind? The answer is that, by use of ingenious and almost magic trick, chemists can cause this same reaction to occur at the flip of a switch by placing the metals involved (conductors of the first kind) in parallel with the solutions involved (conductors of the second kind) and amazingly then turn the reaction on or off with a simple switch! How can this be?

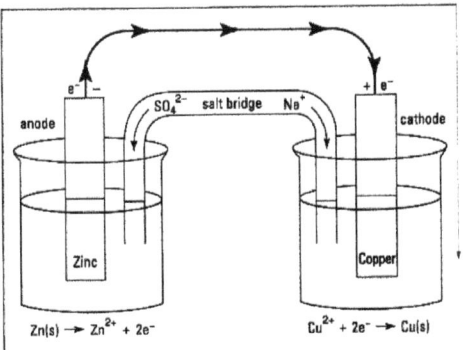

At a rate limited by the ion mobility the electrons on the left can now proceed through the wire and enter the right electrode The process of charge flow can resume as ions shift through the salt bridge and thus smooth the charge distribution within the conductor of the second kind, that is to say in the solutions.

Now you must remember that ions in water move much more slowly than electrons in wires. As a result the ion motions in the solutions control the rate of flow therefore resulting in slow electrons movement in the wire that has been placed across the cell terminal. In "all metal" circuits electron flow is so rapid it almost seems instantaneous, but here it is slow , limited by the pace of ions swimming around in water.

Chemists call this particular electrochemical cell is called a "Daniell Cell" in honor of John Frederick Daniell, a British chemist and Meteorologist in 1856. It appears in General Chemistry textbooks all over the world but NOT as shown here. This drawing which I found on many internet sites touted as a Daniell cell is, in fact, a run down and therefore pretty useless Daniell cell. Why? It is simple. The artist has connected a wire (conductor of the first kind) from zinc to copper , a sure sign that this cell is "dead.". The cell at its birth, and without this unfortunate wire, had a "slightly over one volt" electrical potential between the electrodes labeled "minus" and "plus." Unfortunately the artist has "shorted out" this cell and thus it will measure zero volts on any voltmeter. What shoud the artist have done? He or she should have **omitted this wire** leaving **empty space** between the two "terminals." Then the mobile electrons would have to stop and wait on the left until the user decided to make use of them. In use something such as a resistor or your cell phone should be drawn as connected to these minus and plus terminals. As soon as your cell phone is turned on these waiting electrons willeagerly course through your cell phone wires and enable you to phone home.this happens electrons will flow from minus to plus and enable you to phone home. Shame on the artist for shorting out your cell phone battery.

Look at the arrangement of zinc, copper and several solutions of these ions in the seemingly complicated electrochemical cell (lower picture). Notice that a semicircle of conductors of the first kind is dipped into a odd container holding at least three electrolyte solutions, all of which are conductors of the second kind, that is slow movers. Will anything happen? Yes it will! Say, this is getting exciting is it not? Here is what will slowly happen:

The zinc electrode desperately wants to give up its electrons and zoom them through the connecting wire to the copper electrode until they reach the copper containing solution. Then they can do their job of making copper ions into coper atoms and therefore deposit additional copper metal onto the copper electrode. But this cannot happen unless the zinc electrode agrees to give up its electrons and leave behind un compensated Zn^{2+} ions. No problem. The zinc ions formed simply go into solution and all concerned would seem satisfied. Ah, but one thing is wrong. There is now too many zinc ions in one place and too few copper ions which are a considerable distance apart.in another part of the cell. This charge inequality would stop the electrons from making their trip. But ions are mobile, though not nearly as mobile as are electrons and thus the electron flow anticipated through the wire has to wait for ions to shift. This means that the ion shifts control the rate at which the electron-ion loop can proceed.

At a rate limited by theion mobility the electrons on the left can now proceed through the wire and enter the right electrodeThe process of charge flow can resume as ions shift through the salt bridge and thus smooth the charge distribution within the conductor of the second kind, that is to say in the solutions.

Now you must remember that ions in water move much more slowly than electrons in wires. As a result the ion motions in the solutions control the rate of flow therefore resulting in slow electrons movement in the wire that has been placed across the cell terminal. In "all metal" circuits electron flow is so rapid it almost seems instantaneous, but here it is slow , limited by the pace of ions swimming around in water.

Chemists call this particulr electrochemical cell is called a "Daniell Cell" in honor of John Frederick Daniell, a British chemist and Meteorologist in 1856. It appears in General Chemistry textbooks all over the world but NOT as shown here. This drawing which I found on many internet sites touted as a Daniell cell is, in fact, a run down and therefore pretty useless Daniell cell. Why? It is simple. The artist has connected a wire (conductor of the first kind) from zinc to copper , a sure sign that this cell is "dead.". The cell at its birth, and without this unfortunate wire, had a "slightly over one volt" electrical potential between the electrodes labeled "minus" and "plus." Unfortunately the artist has "shorted out" this cell and thus it will measure zero volts using any voltmeter. What shoud the artist have done? He or she should have **omitted this wire** leaving **empty space** between the two "terminals." Then the eager-to-move electrons would have to stop and wait on the left until the user decided to make use of them. In use something such as a resistor or your cell phone should be drawn as connected to these minus and plus terminals. As soon as your cell phone is turned on these waiting electrons will eagerly course through your cell phone wires and enable you to phone home. Shame on the artist for shorting out your cell phone battery. It will be dead before its chemistry can be explained.

Now I direct your attention to the next electrochemical cell drawing which is only slightly different from that of the poorly drawn Daniell cell. This artist chose to place the zinc electrode on the right instead of on the left of the drawing. This has no significance for zinc metal is the electron pusher in both cases. Both copper and the silver are both less active electochemically tivethan is zinc, and as a result the zinc electrode is the electron pusher in both drawings.

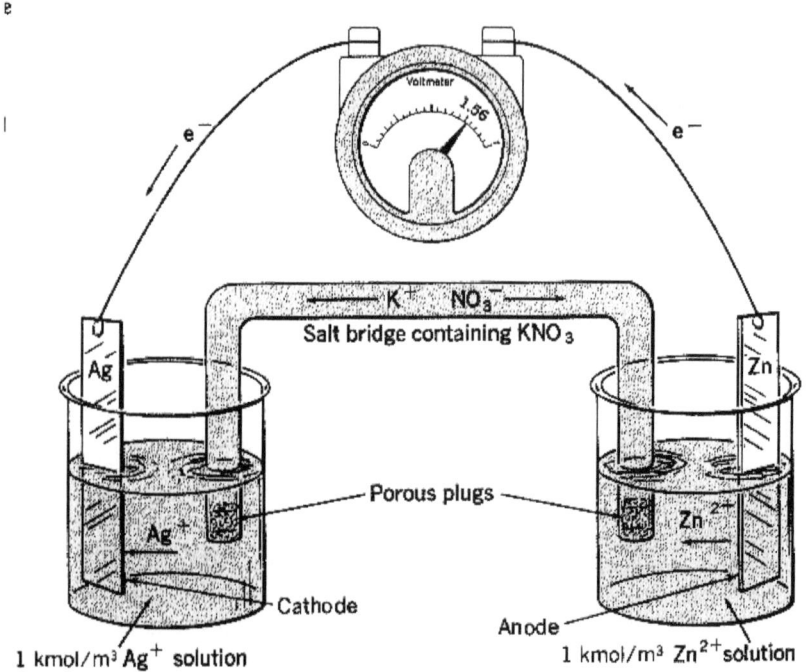

The meaningful feature of this second drawing that is **not** shorted out by a wire and therefore not dead but alive! It has electrons on its protruding zinc electrode waiting for cell phones or other loads to be connected in such a way that these waiting electrons can escape throught this "load" and do work in the real world. This raises the question "what happens to the **zinc ions** left behind in solution? Well when abandoned they disappear into the electrolyte solution and slowly wander off in search of the salt bridge. All of this is slowed down by the slow pace of ions in conductors of the second kind, the ionic solutions. When the cell phone is turned on the zinc ions will begin their journet through the salt bridge and as they go push other positive ions to the copper electrode. At the end of this ion-electron loop Ag+ ions then deposit on the silve electrode, This dance of the electrons, slowed by the chaperone ions,is what makes all electrochemical cells function. Need I point out that in all three diagrams

1. The

e

I

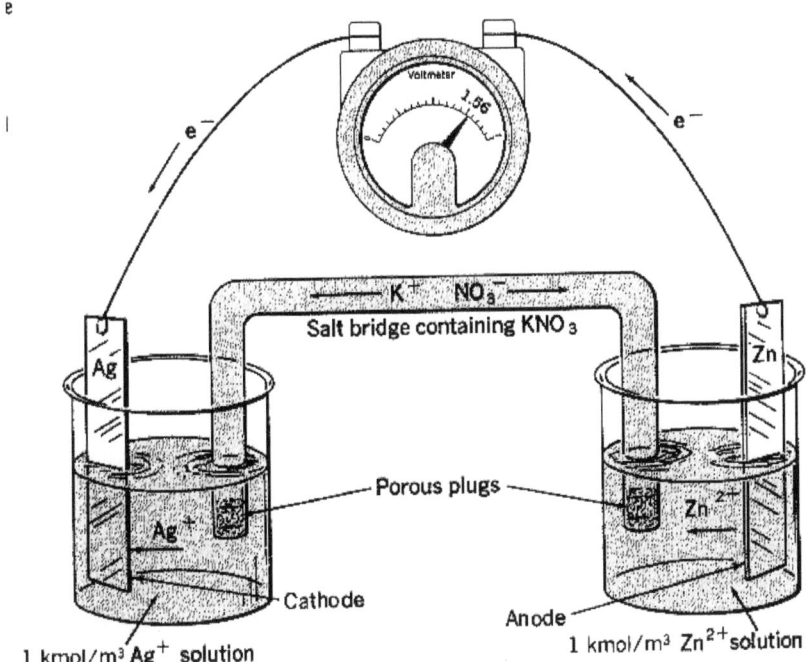

Compare this electrochemical cell with the previous one and you will find that the shorting wire which spoils the previous drawing has been replaced with a **voltmeter**, a device with a very very high electrical resistance and thus the cell delivers **no** current and will not "run down." This cell will have a difference of potential between its terminals, and this difference of potential seems to be 1.56 volts. This circuit because of the very high resistance of the measuring instrument is passing almost zero coulombs of charge. And it tells us numerically that for every coulomb of charge that is permitted to flow (should a something with a lower resistance be connected), there will have been 1.56 Joules of work done as it plates silver on one electrode and erodes zinc from the other.

SUMMARY OF ELECTROCHEMICAL CELLS:

Choose two items listed in the electrochemical series. Build a half-cell in which one half reaction indicated might occur. Provide a metallic wire to communicate electron availability at this wire. Such wires are the vertical lines in most diagrams. Dip this wire into an electrolyte which provides the necessary ions (commonly in H2Osolution).

Build another half-cell choosing a half reaction **higher** in the electrochemical series. This one's wire is to become the negative electrode and of course be dipped on a solution containing the necessary ions. Since it is higher in the series it will offer more electrons for future use in any external circuit. It will be the negative electrode **when the cell is completed**.

The wires sticking out from these two imaginary half cells provide no voltage. Why not? Because voltage is a "difference of potential", also called the Joules per coulomb of charge that can flow **if a circuit** is available. Here the circuit is missing. There is no connection between the two "half-cells" neither top nor bottom. Voltmeters have to connect to **two different points in the same circuit**. Here there is no meaningful "difference of potential, therefore no meaningful voltage to be measured by a voltmeter.

An electrochemical cell is a half-completed circuit. **It becomes an electrochemical cell only when the conductor of the second kind is added at the bottom of the cell diagram.** At last there will be a voltage (difference of potential) between the electrodes (the unconnected wires that stick up).The glassware, and the salt solutions therein, constitute the conductor-of-the-second kind chaperones at this upcoming party. Where's the party? The party begins when conductors of the first kind (your cell phone circuit) is connected to the charged electrodes. When that happens the minus electrode eagerly provide the necessary electrons which end up on the positive electrode and does two things. One, this flow of electrons will eventually neutralise the potential differences and the cell is said to be "run down."

So that is the electrochemical cell. A clever device which offers mobility to electrons once the circuit is completed. The electrons move in a series loop around the circuit if the circuit is complete. If the circuit is not complete electron flow ceases and the cell is "turned off. Reconnecting the circuit resumes this flow of electrons. flowIn many such cells an external D.C. voltage can be applied to its terminals and make it run backward. This process is called "the charging cycle." Is said to return the cell to its original capabilities. Chemists do not like it if you call this cell a "battery." It is not. It is a cell. If several cells are connected in series the package is called a "battery." Because it is a battery of cells. The heavy twelve volt box under the hood of your car is a big fat battery of cells designed in such a way as to provide twelve to thirteen volts (joules per coulomb) at sufficient amperage to "crank the engine" and thus start the car. Then quite cleverly the "generator" in the car recharges the battery enough to permit a return journey.

ELECTROLYSIS

What if you could run a zinc/silver cell backwards? You can by applying an external voltage to the terminals and cause all the half-cell reactions to run backward, un-plating the silver copper and re-plating the zinc? You can. If you have the requisite voltage source and try this zinc will plate on the zinc electrode and copper will un-plate from the copper electrode and urge active metals higher in the electromotive series to come out of hiding and expose themselves in pure metallic form for example sodium and potassium. Humphry Davy and Michael Faraday discovered the elements sodium and potassium in a similar way by passing electric currents through molten salts (water solutions won't work) thus liberating these previously unknown elements to be examined in pure form. The amazing thing is that this was done in the 18th century before the connection between electricity and atoms was a mystery. CONCLUSION: Electroplating and molten salt electrodepositions are not as easy to understand as are familiar electrolysis experiments as running a Daniell cell backwards. The obvious difference is that one might inject electrons into the circuit rather than obtaining them from the circuit.

Actually this would produce deposition of zinc on the eroded zinc electrode and would remove the copper that had thereon been deposited. Surprise surprise! This is exactly what recharging the cell would do.by putting electrons back where they belong. Although this is electrolysis and because recharging of cells (or batteries of cells) needs no discussion here. Therefore my electrolysis lesson is simply to show two examples of electrolysis. The first is a common and easy to understand electrolysis of water on a laboratory scale and the second is an industrial scale electrolysis designed to produce pure aluminium, a large scale example of electron pushing.

So what is **electrolysis**? It is connecting an outside source of voltage from the power company and pushing it into a cell-like arrangement in order to force nonspontaneous reactions to occur. This chapter would be much too long if I gave examples of common electrolysis situations both in laboratory settings and in commercial production. Therefore I show two figures of electrolytic preparations, one a simple laboratory operation, and the other a complex commercial operation.

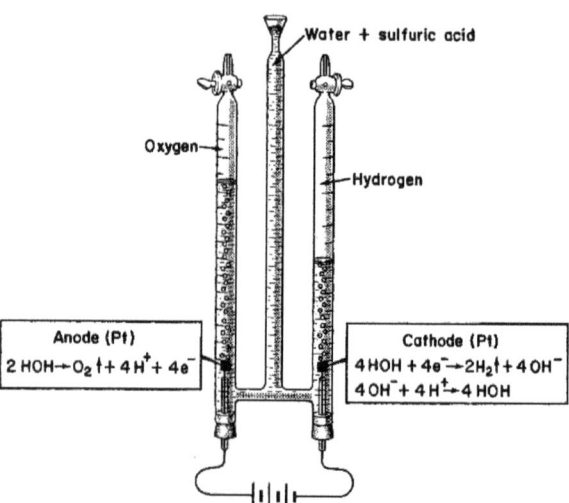

Figure 20-1. Electrolysis of water.

Carbon, coke, or anthracite coal, when heated to dull red, reacts with steam to form a 1:1 molar mixture of hydrogen and carbon monoxide.

$$C + H_2O \rightarrow CO + H_2$$

This mixture is called *water gas* and is one of the commercial methods of producing hydrogen. The carbon monoxide is converted into carbon dioxide by mixing the *water gas* with more steam and passing the mixture over a catalyst.

$$CO + H_2 + H_2O \xrightarrow{catalyst} CO_2 + 2H_2$$

The carbon dioxide is separated from the hydrogen by passing the mixture under pressure through cold water. Carbon dioxide dissolves under these conditions, whereas hydrogen does not.

The reduction of water may be accomplished conveniently by the *electrolysis* of water to which a small amount of sulfuric acid has been added. Hydrogen is reduced at the cathode. Oxidation of water occurs at the anode. The electrode reactions are:

Cathode reduction: $2H^+ + 2e \rightarrow H_2$
Anode oxidation: $2H_2O \rightarrow O_2 + 4H^+ + 4e$
Over-all reaction: $2H_2O \xrightarrow{elec} O_2 + 2H_2$

Reduction of Alumina. Purified alumina is reduced electrolytically in carbon-lined steel tanks. Molten cryolite (Na_3AlF_6) or a synthetic cryolite of NaF, AlF_3, and CaF_2, is used both as the electrolyte and to dissolve the alumina.

A pot line (100 cells connected in a series) operates as a unit at 600 volts with a current of 35,000 amperes. Resistance to the current keeps the cryolite molten at a temperature of about 1900°F.

Aluminum ions are reduced at the carbon lining which serves as the cell cathode. Oxide ions are oxidized at the carbon anodes. As the oxygen is formed, it combines with the carbon of the anode to form carbon dioxide. Equations for the electrolysis are:

Cathode reduction: $Al^{3+} + 3e^- \rightarrow Al$
Anode oxidation: $2O^{2-} + C \rightarrow CO_2 + 4e^-$

The molten aluminum collects on the bottom of the cell and is periodically withdrawn and cast into various molds.

Two pounds of alumina, 0.6 pounds of carbon, and about 9 kilowatt hours of electricity are required to produce one pound of elemental aluminum.

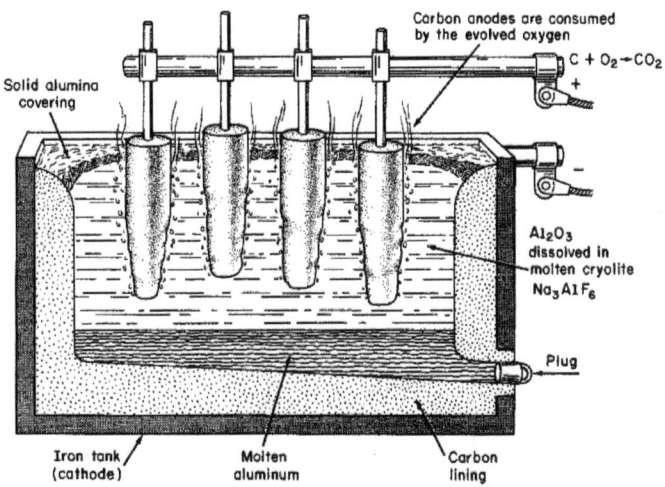

Figure 20-9A. Electrolytic reduction of alumina.

In these examples see if you can tell that electrolysis is the reverse of an electrochemical cell "running down."

Don't strain your brain. I'll simply tell you the answer. When an electrochemical cell is doing its job energy derived from a chemical reaction is being exported to the real world. In electrolysis the reverse is happening. Energy is being expended by the real world to accomplish a chemical reaction that won't go otherwise. I now abandon "cells" to bring up a largely unknown and ignored journey of the separated charges.

ELECTRONS THAT LEAVE HOME AND NEVER COME BACK:

ELECTRETS

Normally this would be the end of this chapter, but in this book I wish to tell you about another little known but related device called an **electret**. An electret is frequently a wax disc that has been specially treated to maintain forever (well almost forever) a plus electric charge on one face and a minus electric charge on the other. A favorite type of wax for such experiments is Carnauba wax .Carnauba wax is wax from a Brazilian tree that is often used to coat candy so that it will not melt. Carnauba wax is a complex mixture of compounds consisting mainly of aliphatic esters (wax esters), α-hydroxyl esters and cinnamic aliphatic diesters obtained from the Brazilian Mart wax palm, *Coperniciacerifera*.

To make an electret take a carnauba wax disc and place a metal electrode on the bottom and an identical metal electrode on the top. You now have a carnauba wax sandwich. Carnauba wax is potentially a conductor of the second kind (ions available) but the ions cannot move (at room temperature). Thus connecting this disc to a high voltage power supply doesn't do much of anything. The ions in it are trapped by their organic waxy lattice.

To **charge** the electret attach it to a high voltage D.C. power supply (say 500 volts) and heat the disc to hot **but not melting**. Now go home and read a good book. Return in the morning and disassemble the whole apparatus. Behold you have created an electret, a dielectric with plus charges on one face and minus charges on the other. This is an **electret**. So what does it do? It stays positive on one side (called the optimist side) and negative on the other, the pessimist side. How do we explain that? One way is to imagine ions embedded in the wax normally staid and immobile can, in a high voltage field, move slowly toward their oppositely charged electrode thus separating ions that were neutralized before the heating. This explains the electret's bipolar behavior. Upon cooling the ions have no chance to return home and an electret is born.

Often (on the internet) the electret's long-lasting electric field is confused with a bar magnet's magnet field. This is wrong. The creation of a magnetic field has **no** commonality with the field of an electret. The electric field creation of the electret may help students who have difficulty understanding polar molecules. A polar molecule is a two atom electret.

Finally the reader may be asking what electret theory has to do with the previous discussion of electrochemical cells. The answer is this: In both electrolysis and cell discharge electrons have to leave home to find work. In electrets the electrons sit there in the wax and never go home.

Chapter 9 RUDY WECKERING

"Scientific theories ...begin as imaginative constructions. They begin, if you like, as stories, and the purpose of the critical or rectifying episode in scientific reasoning is precisely to find out whether or not these stories are about real life"."

Medawar,
Sir PeterPlato's Republic Science and Literature, Sect 4 (p. 53).

Rudy Weckering was a Belgian "engineer", Sortis de l'Universite de Louvain, who wrote a series of three books describing his **Nodic Field Theory** of the atom. Later in this chapter I will list the names of his three volumes (tomes) he wrote to explain his theory of the construction of atoms, the first in published in 1935, the second in 1940, and the third in 1957. Weckering's three books were descriptions his extension of "Lewis structures into the third dimension. This is something Gilbert Newton Lewis had tried to do when he first conceived of the electron in atoms as static (i.e. unmoving) relative to the nucleus. Lewis' drawings were crude because he was not able to sketch the situation he was imagining which was electrons standing still as they arrayed themselves in a simple 3-D pattern about the nucleus. Lewis did his sketching in 1902 when most other chemists were imagining electrons as planets orbiting a sun, an image that turned out to be false.

In this chapter I will give you, the reader, a very brief sketch of the NODIC FIELD THEORY of Rudolf Weckering and its status in the chemical community in the 2oth^t century. But the description of Weckering's theory will have to wait. My first task is to explain what Rudy Weckering has to do with my Clark-Weckering periodic table of the atomes. In tome I (1935) Weckering drew his first suggestion for a periodic table of atoms. Weckering used "atoms" for single atoms just as I propose in this book. What struck me about this table of Weckering's was that the even numbered atoms were arrayed in a separate scheme from that of the odd numbered atoms. Although my tables do not look like Weckening's, still in all versions of my table is this single idea. *Separate the odds from evens*. This means that my table is not just a folded list but is **two** folded lists conjoined. Thank you for the inspiration engineer Weckering.

Lewis' 1902 efffort at static electrons:

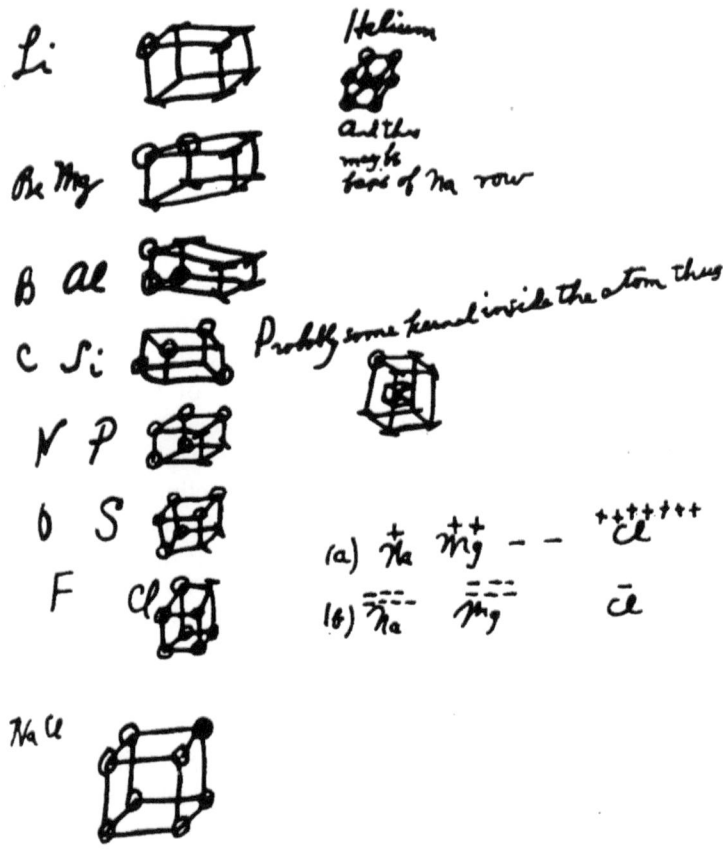

It may take some study of this sketch to realize that Lewis was envisioning a latticework of places were the static electrons were allowed to sit, and since the corners of a cube supplied eight loci the second and third "shell" could be thought of as permitted loci for the electrons of the second and third period. Since this ignored the H and the He electron positions he added a note on the right "some kind of kernel" to provide this. If one imagined these positions as permitted positions for the first eighteen electrons in all atoms, and the electrons were static then Lewis showed how two atoms could share cube edges or cube corners and thus "explain" chemical bonding.

Other chemists took this very seriously and tried to extend Lewis' concept of the static atom. In this author's opinion what made Lewis' 3D static electron ideas fade into history was that no one believed that electrons could stand still at the corners of a cube, and that drawing three dimensional drawing of Lewis' molecules was difficult. In the 1930s Lewis was content for his idea to be dropped except the valence electron count which survives today as a flat octet of dots known today as a "filled electron shell". Thus the Lewis 3D static atom was squashed, yet still very much alive.

Then Inspired by Lewis' static electron drawing and possibly by the ideas of early wave mechanics along came Weckering. This young man knew chemistry, physics and mechanical drawing. Young readers may wonder what "mechanical drawing" is. It is the act of a creative artist sitting at a slanted desk with a few drawing tools and creating seemingly 3D figures such as molecules using only talent and a drawing pen. In the 1930s Weckering took Lewis's abandoned ideas of atoms standing still andredid it with electrons standing still on the nodes of standing waves. In an early quantum idea Weckering placed his electrons on the nodes of the universal standing wave that ruled all atom configurations. With his drawing talent Weckering depicted his electrons in a predicted position in each atom and tied the idea of bonding into this "nodic field theory" .turned them into beautiful drawings of molecules according to the rules of his static electron bonding ideas.

He then drew pictures of bonded atoms (molecules) with size scale and correct bond angles blossoming out of the pages in his "tomes. The results compare well with the computer artistry of today. This began 1935. He then rewrote his Nodic Field Theory in 1942 and then again in 1956."Did he become rich and famous? No. Chemistry ignored Weckering. So how do I know of the nodic field theory? As entered graduate school I received an advertisement for his 1956 book (which is in English.) I was fascinated and have studied his three books since.

A sample Weckering drawing:

Stéréophysique.

Tome II.

STRUCTURE DANS L'ESPACE
DES
EDIFICES MOLÉCULAIRES
DES COMPOSÉS DE LA
CHIMIE ORGANIQUE ACYCLIQUE

Avec 2900 figures dans le texte

par

Rodolphe WECKERING

Ingénieur
Sorti de l'Université de Louvain.

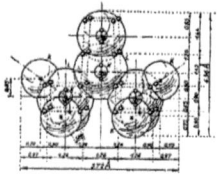

Molécule d'acétone.

PARIS

92, RUE BONAPARTE (VI)

Chapter 6

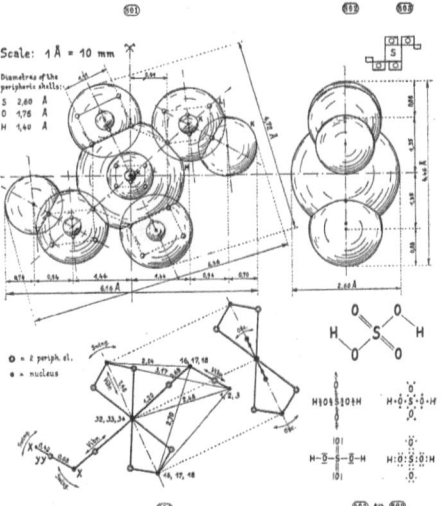

Picture taken from the printed summing up of the author's lecture given at the Congress of Luxemburg of the «Association Française pour l'avancement des Sciences», July 1953, summary of the lectures, page 612. (The dimensions are slightly modified.)

molecule. The indications provided by this sort of model are very helpful for an easier interpretation of the electronic and X-ray interference spectra. The seven masses of the molecule, under the impulse of exterior forces, execute different component movements at sinusoidal regularity, while the linkage electrons are forming the centers of articulations. We distinguish:

a. — **The relative vibration movements** (mouvements vibratoires, Vibrationsbewegungen) of two atoms in the direction XX of the line joining their centers.

b. — **The relative swing movements** (mouvements de basculement, Schaukelbewegungen) in the plane of the molecule.

ATOMIC NODIC FIELD

TWELVE EVIDENTIAL PROOFS, CONFIRMING THE ACCURACY OF THE ATOMIC NODIC THEORY

(The pages and figures refer to the manual, entitled: «The Nodic Field Atom», recently published by the author of the present report.)

A. — Some Fundamental Structural Particarities of the Nodic Field.

A local electro-magnetic field of standing waves, or «**nodic field**», is inseparably connected with each isolated particle, with each atomic nucleus, and with each atomic envelope. See pp. 13, 14, 34, and 97.

This field possesses **electric nodes** (with magnetic vibration bulges), and magnetic nodes (with electric vibration bulges). See p. 15.

In the undeformed field, the nodes are located on concentric spherical shells. See p. 15, and fig. 1 and 2 (p. 14), here reproduced.

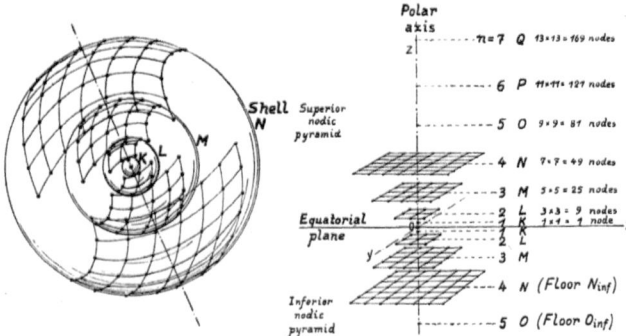

① — Steric structure of the universal electro-magnetic nodic field.
② — Schematized representation of the nodic field.
  (Only the electric nodes are represented.)

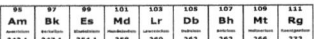

An early Weckering table, inspiration for my odd-even table.

Chapter 10 Chemistry Physics and Energy

"There is a fact, or if you wish, a law governing all natural phenomena that are known to date. There is no known exception to this law – it is exact so far as we know. The law is called the conservation of energy It states that there is a certain quantity, which we call "energy," that does not change in the manifold changes that nature undergoes. That is a most abstract idea, because it is a mathematical principle; it says there is a numerical quantity which does not change when something happens It is not a description of a mechanism, or anything concrete; it is a strange fact that when we calculate some number and when we finish watching nature go through her tricks and calculate the number again, it is the same. **It is important to realize that in physics today, we have no knowledge of what energy "is."** *We do not have a picture that energy comes in little blobs of a definite amount. It is not that way. It is an abstract thing in that it does not tell us the mechanism or the reason for the various formulas."* (italics mine)

Richard Feynman U. S. physicist

"We may say that, before Maxwell, Physical Reality, in so far as it was to represent the processes of nature, was thought of as consisting in material particles....Since Maxwell's time , Physical Reality has been thought of as represented by continuous fields,...and not capable of any mechanical interpretation. This change in the conception of reality is the most profound and the most fruitful that physics has experienced since the time of Newton."

A. Einstein, in James Clerk Maxwell,
A Commemorative Volume. Macmillan, 1931.

"When Faraday first proposed electromagnetic fields around 1839, most scientists thought of them as only a useful way to picture electromagnetic forces and not as real physical objects. Then Maxwell and Hertz showed that waves can travel in electromagnetic fields and that light is one example of these waves. So electromagnetic fields were not just a useful fiction; they were physically real, as real as light. The most convincing argument for the reality of electromagnetic fields comes from conservation of energy. Suppose a radio transmitter sends a message (an electromagnetic wave) to a receiver on Mars and that the message's time travel is twenty minutes. Energy must travel from the sender to the receiver because it takes energy for the receiver to respond. Where is the energy during the twenty minutes between sending and receiving? Not in the sender. Not in the receiver, and energy never just vanishes. So it must be in

the space between sender and receiver, in the electromagnetic field."

Art Hobson[2]

Three quotations! What a way to start a chapter. This is a strange chapter. In quotation one Richard Feynman tells us that we really don't know what energy is, despite frequent assurances that it is "the ability to do work." In the second quotation Albert Einstein tells us that nature presents us with fields to study and not material particles. And to top it all off Art Hobson tells us that electromagnetic waves are "physically real," whatever that means. Feynman was definitely not joking when he flat out stated that *(we have no knowledge of what energy "is.")*

This really shocked me. I had initially wanted to be a physicist but got routed into chemistry by a persuasive chemist. Because of this I spent my career reading extensively in both physics and chemistry never dreaming that physics didn't know what energy was. Three physicists and none of them know what energy is! Chemists know , or so I thought, because they treat it as a reactant in their equations, and even thought that diatomic molecules had energy in them that would leak out if they should be divorced.

Here I am, a chemist, finally writing the concluding chapter of a book about chemistry's three worlds. When I began writing this book I planned to This chapter you are reading now was originally conceived to be about this "third world of the nucleus" and what we know, and don't know about it. I was on the verge of giving this world to the physicists until I read these speculations concerning energy. Now I see that the atom is **not** composed only of massive protons and neutrons and almost-not-there electrons. NO! An atom is composed of massive protons and neutrons and almost not-there electrons AND **energy**! Why has science forgotten that there is a fourth component in all substances, yes even in lone atoms? The fourth component, rarely recognized as a component, is **energy**.

Energy is a component of atoms.

Since my retirement I have read approximately thirty five general chemistry textbooks. None of them listed energy as the fourth component of a lone atom. Invariably they said atoms (they meant atoms) contained three "things", protons, neutrons and electrons. These books do not mention the fourth ingredient, energy, because they do not know what energy is. All the definitions of energy suitable in the first world of chemistry seem inappropriate as a form of energy inside the atom. Textbooks tell the reader of "energy levels" inside of atoms much like unusual males and females living in a two-person-per-room hotel.

---

2 Art Hobson, Physics Concepts and Corrections Fifth Edition,

The student is left to figure out that "energy levels" are apartments for the lonely electrons to share. These apartments (orbitals) will only permit two electrons per apartment. Oh yes, I almost forgot. The two electrons must be of different spins. Orbitals are forbidden to homospin couples. It follows that if there are **energy levels** in the atome there must be **energy** in the atome.

These apartments for electrons are called "energy levels". The electrons that live there are not doomed to live in the same apartment forever. If someone sends them a packet of energy they are permitted to climb to higher unoccupied apartments so that they may see what is going on in the world around them. It has been my experience as a chemist that this promotion to a higher level is a temporary thing. The way electrons normally behave is that they throw away the packet of energy and fall back down the stairs to their initial homes thereby restoring the original atome. But not always. There are daring atomes whose outer electrons actually leap from one atom to another therefore creating atoms of opposite charge. The resulting atomes are then called "ions" and they live together in large quantities in communities called "crystal."

Most atoms though wouldn't be caught dead in such ionic families. Instead they arrange popular and massive dance parties in which they do nothing more shocking than hold hands. These atoms are said to be "covalent." All of this leaping and sharing of electrons is known as "chemical bonding". Why do some atoms behave collectively as bonders while others prefer bachelorhood? It is chemistry's goal to make sense out of all these sharing's and ignoring's of one atom for another. That is what chemists do. What is not always made clear is that "the conditions", also called T and P, are all-important. Chemistry teachers prior to the 20th century taught the chemistry of Earth elements, for that is the chemistry of STP. That is why early periodic charts were charts were **not** charts of **atomes** but were charts of **bonded** atomes which are commonly called atom. At the turn of the century **atomes** (at STP) were discovered and though quite different from bonded atoms, **continued to be called atoms**, a choice which turned out to be a major blunder. To put this another way, free atoms are atomes. Bonded atoms are atoms.

Returning to the topic of "Do atomes contain energy inside of their confines?" I am casting my vote for "yes." As a result it seems logical to ask questions about the mechanism by which energy escapes from atomes and how it gets back in. And if it does get back in what of the many forms of energy does it take on? I am going to delay the answers about energy and matter until Chapter 11, whereupon I shall explain everything. Before I do I would like to quote from Richard Feynman as he addressed a group of teachers about the nature of science and the nature of his father. Relating the teaching his parents, particularly his father, gave him as a youth, Feynman said this:

Begin quotation

*"You might wonder what he (his **father**) got out of it all. I went to MIT. I went to Princeton. I came home, and he said, "Now you've got a science education. I have always wanted to know something that I have never understood, and so, my son, I want you to explain it to me".*

I said yes.

He said, "I understand that they say that light is emitted from an atom when it goes from one state to another, from an excited state to a state of lower energy.

I said, "That's right."

"And light is a kind of particle, a photon, I think they call it."

"Yes."

"So if the photon comes out of the atom when it goes from the excited to the lower state, the photon must have been in the atom in the excited state."

I said, "Well, no."

He said, "Well, how do you look at it so you can think of a particle photon coming out without it having been in there in the excited state?"

I thought a few minutes, and I said, "I'm sorry; I don't know. I can't explain it to you."

He was very disappointed after all these years and years of trying to teach me something that it came out with such poor results."

End of quotation

Why is this Feynman quotation appropriate here? Because "photon is a packet of energy". How can energy exit an atom if energy was not a resident of the atom prior to its emission? At this time in his life young Feynman had taken instruction in quantum mechanics and perhaps even knew the answer to this question which lies hidden in quantum theory, but did not attempt to explain this to his father. It was an excellent question however, and in our 20th century world this question and many others in chemistry have been answered by quantum theory. At the time he had learned the basics of quantum theory but his mastery of quantum theory was to come later. What is quantum theory and what does it have with passage of photons through the walls of atoms? The cell walls in this case are thought of as **mass** and the trapped photon as a blob of pure energy. Questions like electromagnetic waves passing through cell walls can be answered by assuming **that mass evolved from radiation. Mass then is compacted form of radiation. Not only that but mass will someday return to radiation (waves) as it gradually comes unglued.**

Chapter 11 From Energy Fields To Atomes

"Science forces us to create new ideas, new theories. Their aim is to break down the wall of contradictions which frequently blocks the way of scientific progress. All the essential ideas in science were born in a dramatic conflict between reality and our attempts at understanding."

Albert Einstein In Michele Besso,
Correspondence 1903-1955

"A new type of field called a matter field exists in nature. Like EM fields, matter fields are quantized. For example, the matter field for electrons is allowed to possess enough energy for either 0 electrons, or 1 electron, or two electrons, and so on. Electrons (and other material particles) exist because matter fields are quantised in just these energy increments."

Art Hobson...
In Physics Concepts & Connections. Edition five.

"Metaphor is one of our most important tools for trying to comprehend partially what cannot be comprehended totally: Our feelings, aesthetic experiences, moral practices and spiritual awareness. These endeavours of the imagination are not devoid of rationality since they use metaphor, they employ an imaginative rationality."

George Lackoff and Mark Johnson
In Humans as Symbolic Creatures, p 124

"Trying to understand the way that nature works involves a most terrible test of human reasoning ability. It involves subtle trickery, beautiful tightropes of logic on which one has to walk in order not to make a mistake in predicting what will happen. The quantum mechanical and the relativity ideas are examples of this.

Richard Feynman
In The Meaning Of It AllAddison-Wesley

Let us begin this last chapter:

This last chapter is not an attempt to teach quantum mechanics. It is an attempt to present to the reader a presentation of quantum mechanical ideas concerning the relationship between **fields** and **matter**. These widely accepted quantum mechanical ideas are today called "quantum electrodynamics." I do not have the mental capacity at eighty five to claim that I understand what I am going to tell you about this field of endeavour, but I hope that I am able to to explain to my chemistry friends what QM SUGGESTS about the relationship that exists between **fields and matter**. This theory is called "quantum field theory", and I have no interest in deciding whether or not it is right or wrong. but it seems reasonable to me that all the chemistry and physics I learned is not incompatible with these basic ideas of this **quantum field theory**.

- In the beginning there was no matter, only fields. What are fields? From your chemistry background you are familiar with magnetic fields. Radiation fields that you may be familiar with are radio waves, microwaves, television waves, infrared frequencies, lasers, sunlight, ultraviolet light, X-rays, gamma rays, and other waves that we know not of.
- Then **time** came along and suggested it might be amusing if these waves could have wild parties and so they did. (You can't have dances without time.) Waves began to go to their favourite parties and to fill space with music (the sound of music?) until each dancing place was called an electron. Electron dancing places are complex waves localized in one small region. These regions of space we today called "electrons." Is it any surprise that electrons had a wave as well as a particle personality?
- As the party craze got out of control larger particles were formed. This eventually formed atom parties and even larger groupings. *Just how big might wave dancing get?*

The answer came in 1928 when results of an experiment designed to study electrons was that scientists on a little planet called Earth when studying electrons, found that electrons led a dual life. Sometimes they behaved like particles (black dots) and other times they behaved like waves (squiggly lines). Credit for this discovery I usually given to Louis de Broglie and if you prefer a more scientific explanation of why we believe that electrons have a wave nature **and** a particle nature I recommend this: "The wave nature of the electron" Nobel Lecture, December 12, 1929. This is readily available on the internet.

Back to quantum field theory: I am not in this last chapter attempting to teach quantum field theory. I have learned (in my retirement) enough about it to tell you, my reader, what some of the conclusions of this well-founded theory of fields and matter seem to be.

- In the beginning there was fields. The word "field" here means waves of all kinds. That includes the familiar electromagnetic spectrum of waves and any other waves that might exist today or

have existed "at the beginning", whenever that may have been. The concept of "fields" implies time so I suppose there was a beginning time and a now time.
- In the beginning there were no particles, only fields. Then the field dancing craze began and sufficient field parties (explained above) became so large they condensed into "the smallest piece." Today we have a name for that smallest piece, an electron. Is it surprising that electrons have a wave nature?
- There's nowhere to go but up.
- And so we can guess that even though the little creatures we named electrons have a wave origin it is a good bet that larger particles like protons and neutrons also have a wave origin. Having admitted that we must admit thar atoms themselves are wave concoctions, and since we are made from energy and matter we are part of the wave family along with all of our surroundings. Tne next time you see a human being wave please.
- And so if waves can make a world, which came first? Waves or us? Not having ant reliable knowledge of the answer to this question I offer this guess.
- Waves came first. Out of the available waves the Earth and a few other items were created. Since you are one the most remarkable items created you are able to read this book. Even more remarkably, I am able to write it.

What does the future of this world of waves seem to be? I have bad news. Waves are going back where they came from. Obvious clues are entropy increase and radioactivity.

But do not worry. You still have time to buy this book.

[i] Gensler, W. J. *J. Chem. Educ.* **1970**, *47*, 154.

[ii] Brescia, F. A., John; Meislich, Herbert; Turk, Amos *General Chemistry*; Harcourt Brace Javonovich: New York, 1980.

[iii] Masterton, W. L., Slowinski, Emil J, Stanitski, Conrad L. *Chemical Principles*; 6 ed.; Saunders College Publishing: Philadelphia, 1985.

www.ingramcontent.com/pod-product-compliance
Ingram Content Group UK Ltd.
Pitfield, Milton Keynes, MK11 3LW, UK
UKHW042004230426
12048UKWH00009B/542